SpringerBriefs in Pharmaceutical Science & Drug Development

Arnab De

Application of Peptide-Based Prodrug Chemistry in Drug Development

With special thanks to Prof. Richard DiMarchi
(Standiford H. Cox Professor of Chemistry
and the Linda & Jack Gill Chair in Biomolecular
Sciences at Indiana University, Bloomington)

With foreword by Prof. Jean Martinez
(Legion d'Honneur and Chairman of
European Peptide Society, 2002–2010)

 Springer

Arnab De
Hammer Health Sciences Center
Department of Microbiology
Columbia University Medical Center
New York, NY
USA

ISSN 1864-8118   ISSN 1864-8126   (electronic)
ISBN 978-1-4614-4874-7   ISBN 978-1-4614-4875-4   (eBook)
DOI 10.1007/978-1-4614-4875-4
Springer New York Heidelberg Dordrecht London

Library of Congress Control Number: 2012943626

Printed on acid-free paper

Springer is part of Springer Science+Business Media (www.springer.com)

Foreword

Peptides are known to regulate most physiological processes, serving as endocrine signals, neurotransmitters, or growth factors. Hence, they could be used therapeutically under diverse circumstances. Indeed, peptide-based drugs could potentially be highly effective medicines and they are being increasingly used as drugs in allergy, asthma, arthritis, cancer, diabetes, etc.

Traditionally, small molecules have been used in the pharmaceutical industry by virtue of its oral availability, ability to cross membranes, and inexpensive synthetic procedure. However, while the peptide is much larger and more expensive to synthesize, it could be much more potent, specific, and pose fewer side effects. However a major challenge in their therapeutic use is their relatively short duration of action (measured in minutes). Thus, peptides represent a rich natural source of potential medicines with one notable pharmaceutical limitation being their relatively short duration of action.

A prototypical example of such a peptide is the endogenous hormone, glucagon-like peptide-1 (GLP1) which has a physiological half-life of around 2 min. Immediately after a meal, it stimulates the release of insulin, suppresses glucagon levels, and delays gastric emptying. Since the insulinotropic action of GLP1 is glucose-dependent, it acts only after a meal and not in the fasting state, thus protecting against hypoglycemia. Multiple pharmaceutical companies are therefore exploring the possible use of GLP1 as therapy against diabetes. Longer acting peptide agonists of the GLP1 receptor are also under clinical trial.

In this book, Arnab De, who is currently at Columbia University Medical Center, provides insight into how prodrugs of peptides could be designed to improve the pharmacodynamics of peptide drugs, synthetic procedures to make these prodrugs and bioassays to examine the conversion of the prodrug into the drug under therapeutic conditions. This book is partially inspired by the research conducted by him at Indiana University, Bloomington. He presented this research at the American Peptide Symposium 2009 in Bloomington and was awarded the Young Investigators' Award. This volume provides examples of peptide-based prodrug chemistry using GLP1 as a template. The author illustrates how to design prodrugs that slowly convert to the parent drug at physiological conditions of 37 °C and pH 7.2 driven by their inherent chemical instability without the need for any enzymatic cleavage. I recommend the book to the student community at large and especially to advanced professionals in the field of pharmacy, medicinal chemistry, medicine, and bio-chemistry. A detailed bibliography at the end will allow the readers to consult the primary literature if required.

Prof. Jean Martinez
Chevalier of the Legion d'Honneur
Professor of Chemistry and Medicinal Chemistry,
Vice-Chairman of the Board of Governors of the
University of Montpellier, France
Director of Institut Des Biomolecules Max Mousseron
Chairman of European Peptide Society, 2002–2010

Preface

Macromolecular (specifically peptide-based) drugs could potentially be highly effective medicines. However, they have a relatively short duration of action and variable therapeutic index. An example of such a peptide is Glucagon-like Peptide I which could potentially be used as a revolutionary drug for diabetes. This is because it stimulates insulin only when the blood glucose level is high thereby reducing the risk of hypoglycemia (a significant disadvantage of using insulin is that an insulin overdose is the single most potent cause of life-threatening hypoglycemia). However its short duration of action (half-life of 2 min in plasma) precludes its therapeutic use.

In this volume, the use of novel therapeutics like GLP1 as an alternative to tradition insulin-based drugs in diabetes is described. *Application of Peptide-Based Prodrug Chemistry in Drug Development* elucidates the traditional concept of prodrugs as "specialized non-toxic protective groups used in a transient manner to alter or to eliminate certain limiting properties in the parent small molecule" (IUPAC definition). It goes on to provide insight into how prodrugs of peptides (with GLP1 as an example) could be appropriately used to extend the biological half-life, broaden the therapeutic index of macromolecules, and improve the pharmacodynamics of such drugs. The author explains the logic behind designing peptide prodrugs, synthetic procedures and bioassays to examine the conversion of the prodrug to the drug under therapeutic conditions. The prodrugs described slowly convert to the parent drug at physiological conditions of 37 °C and pH 7.2 driven by their inherent chemical instability without the need for any enzymatic cleavage. The diketopiperazine and diketomorpholine (DKP and DMP) strategies for prodrug conversion are demonstrated in detail with special emphasis on the chemical flexibility that it offers to develop prodrugs with variable time actions.

This book will be of use to chemists, biochemists, medicinal chemists, biologists, and people in the medical profession (doctors). It may be used in undergraduate classes but will certainly help post-graduate students and advanced professionals.

The author is grateful to Prof. Richard DiMarchi (Standiford H. Cox Professor of Chemistry and the Linda & Jack Gill Chair in Biomolecular Sciences at

Indiana University) for valuable suggestions. The foreword for the book has been written by Prof. Jean Martinez (Legion d'Honneur awarded by the French Republic; Professor of Chemistry and Medicinal Chemistry of the University of Montpellier, France; and Chairman of European Peptide Society, 2002–2010).

Acknowledgments

I wish to convey my deepest regards for Prof. Richard DiMarchi at Indiana University, Bloomington for his guidance in everything I do. I also thank Mr. Jay Levy and Mr. David Smiley and Dr. Vasily Gelfanov for their valuable suggestions.

I am also deeply grateful to Dr. Subho Mozumdar for his generous help in different aspects of my life.

I take this opportunity to thank my parents (Dr. Arun Kumar De and Mrs. Manjulika De) for my education, for helping me to imbibe the values that help to get past challenging times. I thank my dear younger brother (Dr. Arka De) for critically reading this manuscript. Finally, this book would not be written but for the patient and invaluable feedback of my dear wife.

Thank you all!

Acknowledgements

Contents

Chapter 1
Introduction

Abstract This chapter introduces diabetes, an ancient disease that continues to affect diverse populations in modern times and the various types of diabetes are discussed. Insulin has been traditionally used to treat the disease. However, the one significant limitation of using insulin is that an insulin overdose is the single most potent cause of life-threatening hypoglycemia. Glucagon-like Peptide-1 (GLP1) is a 30 amino acid peptide that stimulates insulin only when the blood glucose level is high, hence the risk of hypoglycemia is substantially minimized. However, the biggest problem in the therapeutic use of GLP-1 is its extremely short half-life in plasma (~ 2 min). In the last subsection of this chapter, the utility of prodrugs is discussed in general and how GLP-1 based prodrugs could appropriately extend the biological half life and broaden the therapeutic index of GLP.

Keywords Diabetes · Insulin · Glucagon-like Peptide-1 · GLP-1 · Prodrugs · Peptides · Hypoglycemia

1.1 Diabetes

Diabetes is an ancient disease that continues to affect a diverse population in modern times. The first recorded cases of diabetes date to ancient times in Egypt and India (where it was called Madhumeha in ancient Ayurvedic medicine) [1]. The term "diabetes" was first coined by a Greek physician named Aretaeus of Capasdocia [1].

The prevalence of diabetes has grown steadily over the last 30 years, largely as a result of poor diet and a rapid rise in the prevalence of obesity. Diabetes is a disease associated with sizable morbidity and excessive mortality. It imposes an immense financial burden on those afflicted with the disease and general societal

A. De, *Application of Peptide-Based Prodrug Chemistry in Drug Development*,
SpringerBriefs in Pharmaceutical Science & Drug Development,
DOI: 10.1007/978-1-4614-4875-4_1, © The Author(s) 2013

health care costs. It is broadly accepted that there currently is a worldwide epidemic of Type 2 diabetes (often referred to as adult-onset diabetes). Moreover, numerous clinical studies have shown that most cases could be prevented or managed by lifestyle modifications and proper medication [2]. In certain demographic populations, like the Pima Indians of North America, as much as 30 % of the adult population has been diagnosed with Type 2 diabetes. In the United States alone, 18 million people (6.3 %) currently suffer from diabetes which in turn is a leading cause of blindness and heart attacks, as well as kidney and vascular disease [3, 4].

The primary physiological cause of diabetes is the defective utilization of glucose by insulin-responsive cells in the body. Consequently, blood glucose levels increase despite elevations in insulin concentrations and hyperglycemia eventually emerges. The glucose accumulates in the blood instead of being absorbed and metabolized. Therefore, the cells do not generate enough energy to perform their normal activities while the persistently high blood glucose concentration is damaging to numerous tissues, especially the eyes, kidney and nerves.

There are three different forms of diabetes that are distinguished by their etiological onset and progression. The physiological effects vary in severity and cause, but all induce similar types of damage to physical health.

(1) Type 1 diabetes: Type 1 diabetes is often referred to as juvenile diabetes because the majority of cases strike before adulthood. This is a chronic disease of childhood and approximately 150,000 people under the age of 18 and more than a million people in total are afflicted by this disease in the United States alone. Its prevalence is rising at a rate of 3 % a year [5]. Over time, Type 1 diabetes can lead to serious medical complications such as cardiovascular diseases, diabetic retinopathy and diabetic neuropathy [6]. Type 1 diabetes is most often the result of a humoral based auto-immune response against β cells of the islets of Langerhans which are located within the pancreas. These cells are responsible for producing the insulin required for normal metabolic homeostasis. Such patients typically lose about 80–90 % of their β cells [7] and the remaining population is insufficient to meet the body's normal insulin requirements. This leads to hyperglycemia or what is also known as "insulin-dependent diabetes". People diagnosed with Type 1 diabetes need to be treated with daily insulin injections.

Type 1 diabetes also presents us with a challenging paradox. If the subjects could be identified before hyperglycemia occurs, the initiation of the auto-immune process might be prevented, thus halting the development of diabetes. Unfortunately, a definitive diagnostic method to determine who will eventually develop Type 1 diabetes does not exist and attempts to develop such methods would require large numbers of test subjects. Without such a diagnostic, the development of an efficacious therapy is difficult since finding the subjects is very hard and the risk—benefit ratio of an experimental medicine in a subject prior to disease onset is unknown [6].

(2) Type 2 diabetes (non insulin-dependent): This is the more common form of diabetes. A healthy person's body secretes enough insulin to maintain a steady blood glucose level. In Type 2 diabetes, the body does not produce enough insulin in a relative and often an absolute sense. Resistance to insulin action is a physiological hallmark feature of this disease. This resistance to insulin is often caused by obesity [8]. Although this is not a universal phenomenon, underweight patients are often found to have had impaired insulin secretion while the obese exhibit "insulin resistance" [9].

Approximately 90–95 % patients that suffer from diabetes reportedly have Type 2 diabetes [10] which normally occurs after 40 years of age. Hence it is known as "adult-onset diabetes". Due to changes in dietary habits and lack of exercise, it is no longer uncommon for Type 2 diabetes to occur in younger people, even adolescents. This shows a strong correlation with the alarming rise in the prevalence of obesity and a sedentary lifestyle.

(3) Gestational diabetes: This occurs in pregnant women, hence the name. If the blood glucose level is high in a woman during pregnancy but not at other times, then she is said to have gestational diabetes. The cause is currently unknown but seems similar to Type 2 diabetes where pregnancy imposes a level of insulin resistance. Approximately 4 % of pregnant women are purported to have this disease, and around 135,000 new cases are reported every year [11]. Further observations show that approximately 50 % of gestational diabetes reappears as Type 2 diabetes within 2 years of child bearing [11].

1.2 Insulin

In two landmark papers [12, 13] in 1922, Fredrick Banting, Charles Best, James Collip and John Macleod reported the extraction of insulin from the pancreas of a dog. This extract was subsequently shown to lower the blood glucose level in surgically induced diabetic animal models. The clear demonstration that diabetes is caused by the deficiency of insulin and could be reversed pharmacologically makes these two of the more important scientific papers of the 20th century.

Insulin is a rather small protein, with a molecular weight of slightly less than 6,000 Daltons. The primary amino acid sequencing was accomplished by Sanger in 1959 [14]. It is composed of two peptide chains, designated as the A chain and B chain. The A chain has 21 amino acids while the B chain has 30.

The chains are linked by two disulfide bonds (residues A7 to B7, and A20 to B19). In addition, there is a third intramolecular disulfide bond in the A chain (residues A6 to A11) (Fig. 1.1). The complete conservation of these three disulfide bonds throughout the mammalian phylum underscores the critical importance of this bonding pattern. This also increases the structural complexity of the molecule and makes it more difficult to synthesize as compared to a single peptide chain.

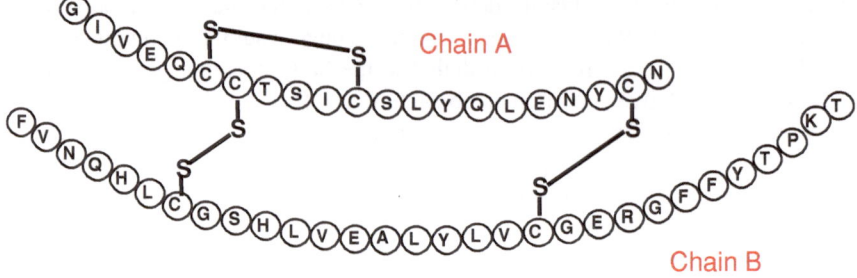

Fig. 1.1 Primary structure of Insulin

Table 1.1 Pharmacodynamics of common insulin analogs

	Onset (h)	Peak Action[a] (h)	Duration[a] (h)
Humalog®	0.25	0.5–1.5	3–5 (very fast)
Regular	0.5	2–4	6–8 (fast)
NPH, Lente	1–3	6–12	18–24 (slow)
Ultralente	4–8	12–18	24–28 (very slow)

[a] Since patients respond to insulin differently, the peak action and duration are given as ranges

There are three important pharmacodynamic characteristics of a drug: its onset (duration to biological action), its peak time (time at which the biological effect is strongest), and finally its duration (sustained time of biological activity). There are several unique forms of insulin designed to meet an individual patient's daily glucose demands (meal and fasting level). They are commonly classified into four broad categories by their duration of action: very fast, fast, slow, and very slow (Table 1.1).

The time action curves of some common insulin analogs are shown in Fig. 1.2. Ultralente is a very slow acting insulin, hence it is usually used with Humalog® (very fast acting) or native regular insulin (fast acting) [15] to more accurately mimic the normal daily physiological variation in insulin activity.

Numerous combinations of these insulin analogs facilitate regulation of blood glucose in virtually all forms of diabetes. However, they all have several severe limitations. The most relevant is the finding that insulin overdose is the single most potent cause of life-threatening hypoglycemia [16]. This has been confirmed by many clinical trials, most notably by the Diabetes Control and Complications Trial Research Group [17]. Studies performed in representative populations clearly demonstrate that weight gain is another problem associated with insulin therapy [18]. Such weight gain can paradoxically increase insulin resistance and thus the amount of insulin needed. Obesity also increases cardiovascular risks [19].

While insulin is a miraculous substance it is a challenging medicine. Its complex structure, especially its intra and intermolecular disulfide bonds makes it difficult to synthesize in an inexpensive manner and the molecule hard to

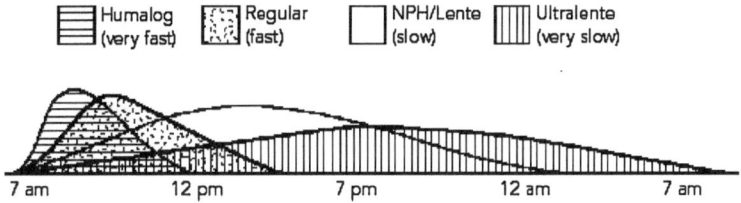

Fig. 1.2 Pharmacodynamics of common insulin analogs

formulate. Furthermore, the natural tendency of insulin to form insoluble multimeric complexes at high concentrations seriously complicates its commercial production. Additionally, the narrow therapeutic index and thermal instability of insulin over extended periods of time makes refrigeration a necessity. Lack of adequate refrigeration is a major issue in many parts of the world.

What is most needed is a medicine that is capable of normalizing blood glucose without the risk of hypoglycemia. In this context, it is helpful to note that Glucagon-like peptide 1 (GLP-1) therapy has been shown to increase native insulin synthesis and secretion without inducing hypoglycemia [20].

1.3 Glucagon-Likepeptide 1 (GLP 1)

Glucagon-like peptide 1 is a hormone that offers the promise of revolutionizing the treatment of Type 2 diabetes. In a landmark paper published in 198 [5, 21] two endogenous peptides were reported which had a high sequence homology to glucagon and, like insulin, displayed high conservation across a range of different species. These two peptides were Glucagon-like peptide-1 (GLP-1) and Glucagon-like peptide-2 (GLP-2) which were first identified in the course of cloning the gene for proglucagon. Upon testing, it was found that GLP-1 stimulated insulin release while GLP-2 did not [21]. Early clinical experiences suggested that GLP-1 has an attractive pharmacologic profile [22]. Throughout this book, GLP-1(7-37)-acid has been denoted simply as GLP with changes added to this nomenclature to signify related peptides.

GLP is secreted from the gut in response to a meal [23]. It is an incretin hormone that has the potential to offer an ideal treatment for Type 2 diabetes [24]. GLP enhances the secretion of insulin [25] only when the blood glucose level is high, eliminating the risk of hypoglycemia. It inhibits glucagon secretion therein maintaining an optimal ratio of insulin and glucagon [26]. GLP has further been shown to reduce food intake via its effects on gut motility (it inhibits the motility of the upper gut), leading to weight loss and decreased obesity [26]. Thus, GLP by virtue of its multiple biological actions has emerged as a valuable tool in the treatment of Type 2 diabetes, and related metabolic syndromes.

Human Glucagon Like Incretin Peptide (7-37)-OH

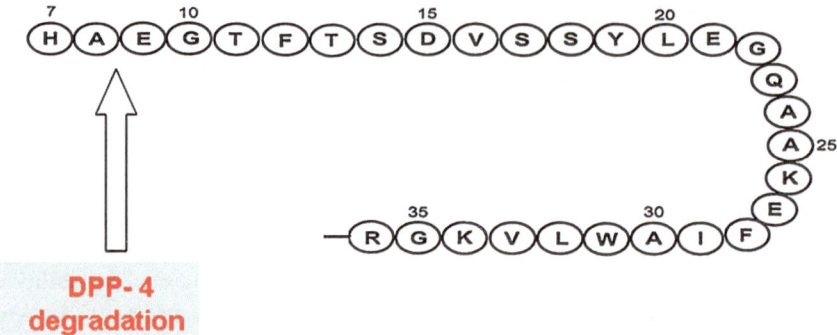

Fig. 1.3 a Primary structure of GLP-1(7–36) amide and action of DPP IV, **b** Primary structure of GLP-1(7–37) acid and action of DPP IV, **c** 2D-NMR structure of GLP-1 The 2D-NMR structure of GLP-1 has revealed that the first 7 residues show a random coil (*pink*) followed by a turn (*yellow*). This is followed by a helical region (*red*; residues 7–14), a linker region (*yellow*; residues 15–17) and another helical region (*red*; residues 18–25). The C-terminal region of GLP-1 is again a random coil. It is speculated that the N-terminal random coil region interacts with the receptor. The structure has been drawn using PDB Protein Workshop software. [34]

There is another more physiological factor to rationalize GLP therapy. Type 2 Diabetes is characterized by insufficient insulin secretion and declining β-cell function. This β-cell defect partly results from the progressive loss in β-cells function, as noted previously. GLP also has an apparent mitogenic effect on the β-cells of the pancreas [27], and thus stimulates the in vivo biosynthesis of insulin (thus addressing the main defect of progressive decay in β-cell capacity) [26]. Additionally, recent research demonstrates that almost two thirds of the insulin secreted in response to a meal is because of the action of insulinotropic actions of the incretin hormones, like GLP [28]. Indeed patients with Type 2 Diabetes have been long known to exhibit a variable loss in the incretin action of GLP [29]. In this way, GLP therapy addresses the physiological replacement for the loss in incretin action associated with Type 2 Diabetes.

GLP stimulates the secretion of insulin by interacting with the GLP G-Protein Coupled Receptor (GPCR) expressed on the surface of β-cells of the pancreas. This receptor is coupled positively to the adenylyl cyclase system [30]. After ligand activation, the adenylyl cyclase is stimulated leading to an increase in cAMP concentration within the β-cells [31]. This in turn activates protein kinase A that leads to an avalanche of additional biochemical events [32]. GLP mediated receptor action via adenylate cyclase is the basis of our in vitro bio-assay, and has proven useful in design of GLP analogs with therapeutic promise (see Experimental Design of Bioassays). It is to be noted that this insulinotropic effect of GLP is strictly glucose dependent [33].

Human Glucagon Like Incretin Peptide (7-37)-OH

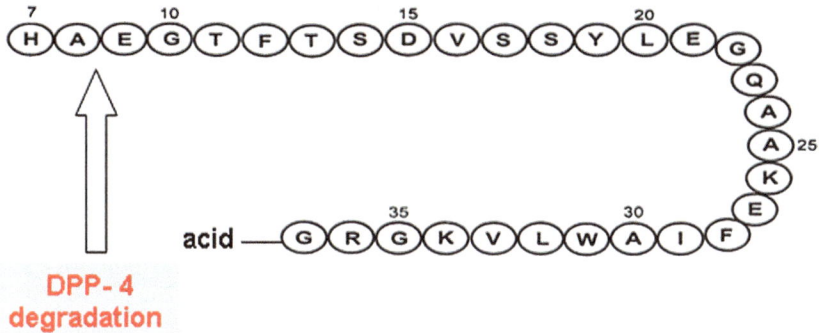

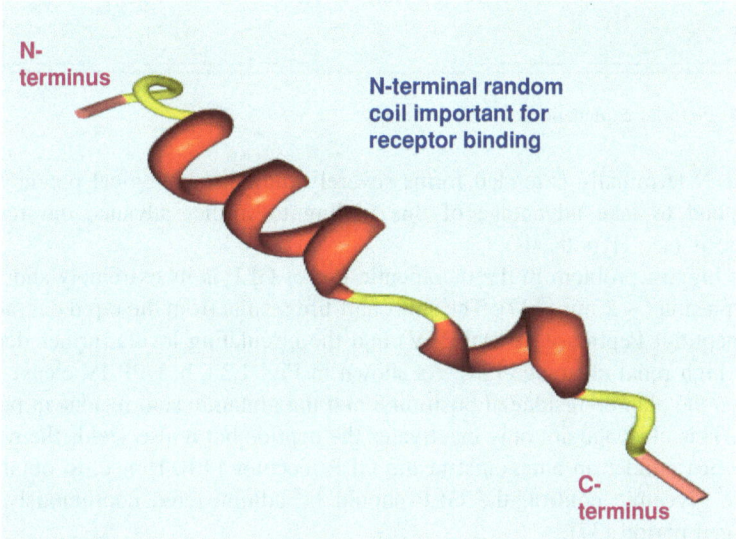

Fig. 1.3 continued

Glucagon-like peptide 1 physiologically exists in two forms of comparable biopotency [35]. The first is the more abundant [35] GLP(7-36) amide where the C terminus is an arginine amide (Fig. 1.3a). The second form is the GLP(7-37) acid where the C terminus is a glycine (Fig. 1.3b). The secondary structure of GLP-1 is shown in Fig. 1.3c.

As illustrated in Fig. 1.3c, it is thought that the N-terminal random coil region is important for receptor binding. Structure–activity relationship studies have shown that the N terminal histidine is especially important for the potency of GLP

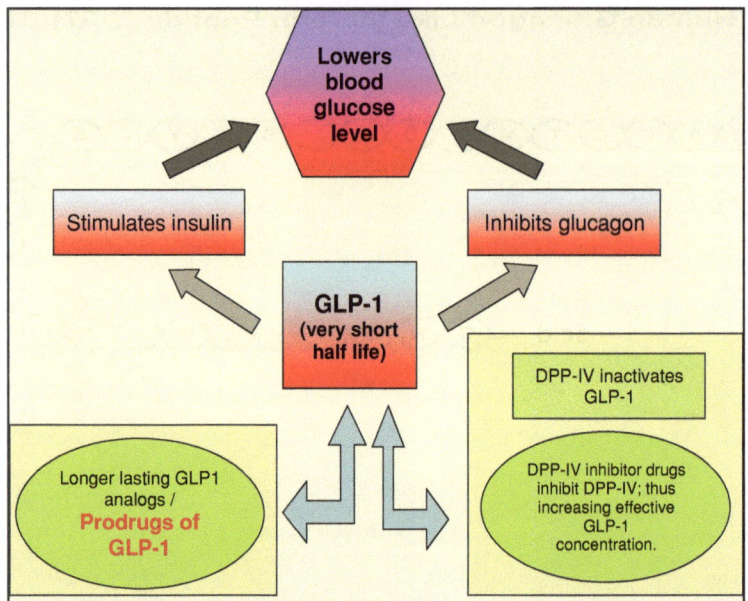

Fig. 1.4 Pictorial explanation of GLP-1 therapy

and the N-terminally extended forms severely diminish biological potency [36]. We intend to take advantage of this finding to further advance out research hypothesis (see Hypothesis).

The biggest problem in the therapeutic use of GLP is its extremely short half-life in plasma (~ 2 min) [37]. The short half life results from the rapid degradation by Dipeptidyl Peptidase IV (DPP-IV) and the circulating levels further decrease due to high renal clearance [38]. As shown in Fig. 1.3a, b, DPP-IV cleaves GLP between the alanine residue at position 8 and the glutamic acid residue at position 9 [39]. This cleavage not only inactivates the peptide but it also yields the residual GLP(9-36) amide; an antagonist at the GLP receptor [40]. Hence, to obtain reasonable glycemic control, the GLP should be administered continuously for a prolonged period [37].

Two other strategies have been formulated to use GLP signaling to treat Type 2 diabetes: use of a DPP-IV inhibitor [41] or design of a longer-acting variant of GLP (pictorially represented in Fig. 1.4).

The most obvious difference between these two therapies seems to be their effect on body weight. While native GLP and its analogs promote loss of weight, the DPP-IV inhibitors act by preventing gain of weight [42]. Additionally, several other issues regarding the feasibility of using DPP-IV inhibitors need to be addressed before effective therapeutic use [43]. Even after treatment with DPP-IV inhibitors it is not clear if there would be enough GLP to attain peak efficacy in the body. This is because the amount of endogenous GLP secreted is severely reduced in diabetes. The fact that DPP-IV non-specifically acts on peptides other than GLP adds to its efficacy but is also a matter of concern. The larger issue pertains to the other useful functions of DPP-IV in different tissues, such that adverse effects of DPP-IV inhibitors are theoretically possible [43].

Our work concentrated on designing a prodrug for GLP i.e., a longer-acting variant of GLP. A few longer-acting analogs of GLP have been identified, two of which will be discussed below. The first peptide is Exendin-4 which is an agonist for GLP receptor [44] and is found in the saliva of the Gila monster [45] (Heloderma species native to several American States). Exendin-4 was discovered by John Eng, an endocrinologist at the Bronx Veteran's Affairs Medical Center in New York. It is being marketed by Amylin Pharmaceuticals. Exendin-4 functions in a manner similar to that of native GLP with which it shares 53 % amino acid homology. Several important functional residues like the N-terminal histidine are conserved. It is much more resistant to cleavage by DPP-IV than GLP as a result of the glycine at the second residue instead of alanine. Thus, it is found that Exendin-4 has all the biological functions of GLP, but with a longer half life (60–90 min) [46]. While the $t_{1/2}$ is much greater as compared to native GLP ($t_{1/2} \sim 2$ min), an even longer-acting drug is highly desirable. Finally, since Exendin-4 is of non-human origin, it has been reported to induce an immune response that compromises long-term clinical efficacy [47].

Another longer-acting GLP analog is called Liraglutide. In this peptide analog, there is an arginine in the 34th position instead of a lysine. Additionally, a glutamic acid residue is added to the lysine residue at position 26 [48] and the α-amine of this glutamate is acylated with a C16 fatty acid. Upon entering the bloodstream, the fatty acid non-covalently binds to serum albumin (a protein facilitating plasma transport) [49]. The serum albumin distributes the drug throughout the body and also protects against cleavage [48]. As a result Liraglutide's $t_{1/2}$ is increased to about $10-14$ h and the drug can be injected once a day. This drug is being developed by Novo Nordisk and has now entered Phase 3 trials.

The N-terminal histidine of GLP (7th position as shown in Fig. 1.3b) is important for pancreatic receptor activation of GLP and exendin-4. Previous amino acid substitutions for this histidine with lysine and alanine have resulted in a large loss of bioactivity [36]. Hence it has been concluded that the imidazole ring of the histidine is necessary for the full potency of the molecule. GLP lowers blood glucose in diabetic patients, and might restore β cell sensitivity to exogenous insulin secretagogues. Studies so far have yielded evidence that GLP therapy is safe and effective for Type 2 diabetics [50]. We want to develop a longer-acting GLP prodrug, possibly with weekly or even monthly duration of in vivo biological action.

1.4 Prodrug

A prodrug is the precursor of a drug. According to The International Union of Pure and Applied Chemistry (IUPAC), the term prodrug is defined as "any compound that undergoes biotransformation before exhibiting its pharmacological effects.

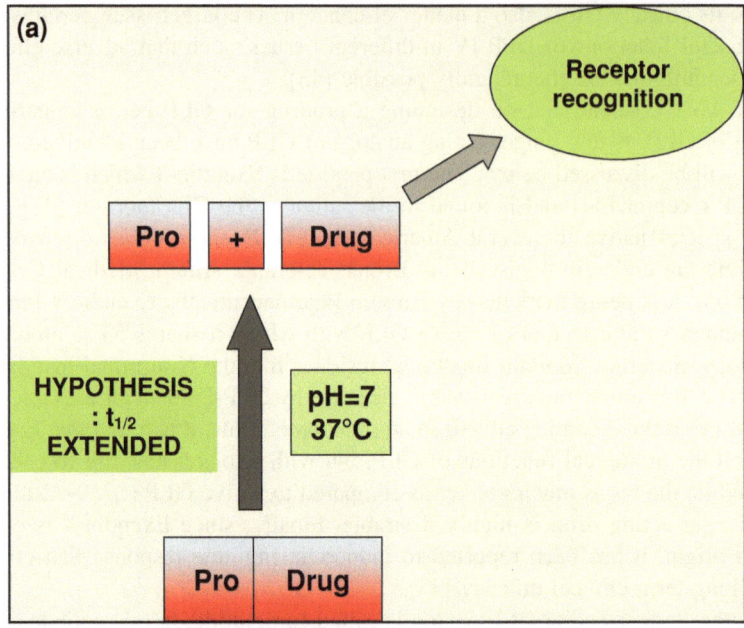

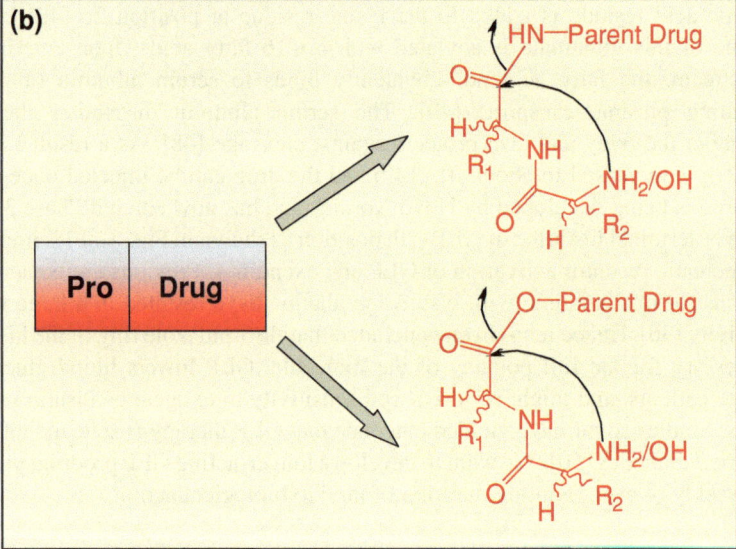

Fig. 1.5 a General prodrug hypothesis, **b** Hypothesis (N-terminal amide and ester prodrug)

Prodrugs can thus be viewed as drugs containing specialized non-toxic protective groups used in a transient manner to alter or to eliminate certain limiting properties in the parent molecule" [51]. The term "prodrug" was first introduced by Albert in 1950 to signify pharmacologically inactive chemical derivatives that undergo conversion endogenously to become an active pharmacological agent to increase their usefulness or decrease their toxicity [52]. In this regard, prodrugs can be used for various purposes. Most prodrugs are designed to facilitate high cellular absorption following administration [53] (i.e. valaciclovir). Prodrugs must be soluble in an aqueous media to be absorbed properly. Insolubility of a drug can also cause significant pain at the site of injection. An example is clindamicyn where injections are painful, but its phosphate ester prodrug improves solubility and alleviates the pain [54]. The prodrug should be non-toxic, stable in storage and must be resistant to degradation in different body fluids until that point when it reaches its site of action. Finally after reaching the specific site of action, there should be a quantitative release of the drug [54].

The prodrug approach is commonly utilized with small molecules (<500 Daltons) [55] to enhance oral delivery. The main obstacle to using peptides as potential drugs is their short half life in the systemic circulation because of proteolytic hydrolysis and the rapid clearance by blood, liver and kidney [56]. A second limitation is their limited oral absorption [56]. Due to these problems of short half life and poor oral bioavailability, alternative routes of delivery are being explored [57].

The prodrug that we envision has certain unique characteristics. Through structural refinement, we intend to appropriately extend the biological half life and broaden the therapeutic index of GLP. Our prodrug concept is not focused on oral bioavailability as with conventional small drug approaches, but upon extended biological half life. Consequently, many of the stringent necessities of conventional prodrugs are not relevant. Most prodrugs as described above are designed to facilitate transport across biological membranes. Our prodrug is used to delay the time of action by inhibiting recognition by the corresponding receptor. Receptor recognition is the primary means of degradation and thus termination of biological activity of our drug. We seek to convert our prodrugs to active peptides by controlling the chemical conversion to structures that can be recognized by the receptor. The speed of this chemical conversion will determine the time of onset and duration of in vivo biological action (Fig. 1.5).

The final element of this work is the application of selective pegylation [58] to delay non-productive, premature, in vivo clearance. Peptides are easily cleared because of their relatively small molecular size when compared to plasma proteins. Increasing the molecular weight of a peptide above 40 kDa exceeds the renal threshold and significantly extends duration in the plasma. The judicious choice in the site of attachment of a polyethylene glycol polymer that is 10-fold larger than insulin or GLP is a sizable challenge.

References

1. Majumdar SK (2001) Bull Indian Inst His Med Hyderabad 31(1): 57–70
2. Pendergrass M (2007) Nat Endocrinol Metabol 3(1):1
3. http://www.diabetes.org

4. Cryer PE (2007) Nat Endocrinol Metabol 3(1):4–5
5. Onkamo P et al (2000) Diabetologia 43(10):1334–1336
6. Kishiyama CM, Chase H, Barker JM (2006) Rev Endocr Metab Disord 7(3):215–224
7. The Diabetes Control and Complications Trial Research Group (1998) Ann Intern Med 128(7):517–523
8. Cavaghan MK et al (2000) J Clin Investig 106:329–333
9. Sadikot SM Jaslok Hospital and Research Centre, Mumbai http://diabetesindia.com/
10. United Kingdom Prospective Diabetes Study (1995) Diabetes 44:1249–1258
11. http://www.diabetes.org/gestational-diabetes.jsp
12. Banting FG et al (1922) Can Med Assoc J 12:141–146
13. Banting FG, Best CHJ (1922) J Lab Clin Med 7:464–472
14. Sanger F (1959) Science 129(3359):1340–1344
15. Lepore M et al (2000) Diabetes 49(12):2142–2148
16. De León DD, Stanley CA (2007) Nat Clin Pract Endocrinol Metab 3:57–68
17. The Diabetes Control and Complications Trial Research Group (1993) The. N Engl J Med 329(14):977–986
18. Nicholas GA, Gomez-Caminero A (2007) Diab, Obes Metab 9(1):96–102
19. Johnston LW, Weinstock RS (2006) Mayo Clin Proc 81(12):1615–1620
20. Nauck MA et al (2002) J Clin Endocrinol & Metab 87(3):1239–1246
21. Schmidt WE, Siegel EG, Creutzfeldt W (1985) Diabetologia 28(9):704–707
22. Green BD et al (2003) Biol Chem 384:1543–1551
23. Holst JJ (1999) Trends Endocrinol Metab 10:229–234
24. McCormack JG (2006) Biochem Soc Trans 34:238–242
25. Orskov C et al (1996) World J Gastroenterol 31:665–670
26. Daniel J et al (2006) Lancet 368(9548):1696–1705
27. Giorgino F et al (2006) Diab Res Clin Pract 74(2):S152–S155
28. Butler AE et al (2003) Diabetes 52:102–110
29. Nauck M et al (1986) Diabetologia 29:46–52
30. Drucker DJ et al (1987) Proc Natl Acad Sci 84:3434–3438
31. Holz GG et al (1999) J Biol Chem 274:14147–14156
32. Gromada J et al (1998) Pflugers Arch 435:583–594
33. Weir GC et al (1989) Diabetes 38:338–342
34. Hui H et al (2005) Diab/Metab Res Rev 21:313–331
35. Rabenhoj L et al (1994) Diabetes 43(4):535–539
36. Adelhorst K et al (1994) J Biol Chem 269(9):6275–6278
37. Larsen J et al (2001) Diab Care 24:1416–1421
38. Ørskov C, Wettergren A, Holst JJ (1993) Diabetes 42:658–661
39. Deacon CF et al (1995) J Clin Endocrinol Metab 80:952–957
40. Knudsen LB, Pridal L (1996) Eur J Pharmacol 318:429–435
41. Holst JJ, Deacon CF (1998) Diabetes 47(11):1663–1670
42. Åhren B et al (2004) J Clin Endocrinol Metab 89:2078–2084
43. Stuart AR, Gulve EA, Minghan W (2004) Chem Rev 104(3):1255–1282
44. Thorens B et al (1993) Diabetes 42(11):1678–1682
45. Eng J et al (1992) J Biol Chem 267(11):7402–7405
46. Kolterman OG et al (2005) Am J Health Syst Pharm 62:173–181
47. Kendall DM et al (2005) Diab Care 28:1083–1091
48. Knudsen LB et al (2000) J Med Chem 43:1664–1669
49. Bjerre KL et al (2005) Diabetes 52(1):321–322
50. Deacon CF (2004) Diabetes 53:2181–2189
51. IUPAC Pure & Applied Chemistry (1998) 70(5):1129–1143
52. Albert A (1958) Nature 182:421–423
53. Jiunn HL, Lu AYH (1997) Pharmacology Rev 49:403–449
54. Stañczak A, Ferra A (2006) Pharmacol Rep 58:599–613
55. Beaumont K et al (2003) Curr Drug Metab 4:461–485

56. Humphrey MJ, Ringrose PS (1986) Drug Metab Rev 17:283–310
57. Saffran M et al (1988) J Pharm Sci 77:33–38
58. Lee SH et al (2005) Bioconjug Chem 16:377–382

Chapter 2
Application of Prodrug Chemistry to GLP-1

Abstract It is impressed upon the reader how a prodrug at the biologically active N-terminal end of GLP could be utilized to extend and improve the pharmacodynamics of this peptide hormone. The quality of an ideal prodrug and the different strategies that could be potentially considered for constructing peptide-based prodrugs are described . One could use enzymes for the conversion of a prodrug to a drug but here we relied on pH and temperature for generating the drug from the prodrug. This is because these two attributes are virtually invariant physiologically and thus the strategy could be widely used. Thiol prodrugs are described, but they are relatively unstable in physiology and thus would likely cleave too fast, thereby resulting in rapid clearance of the prodrug in its active form. Finally, the intramolecular cyclization reaction of dipeptide esters and amides to form diketopiperazine or diketomorpholine is considered as an example of a chemoreversible prodrug. The chemistry is reasonably straightforward and allows at least four points (stereochemistry of substituent groups, nature of the nucleophile and leaving group as shown) where structure can be stereochemically controlled to refine the rate of formation with release of the active peptide.

Keywords Ideal prodrug · Diketopiperazine · DKP · Diketomorpholine · DMP · Active imidazole · Pharmacodynamics · Therapeutic index

2.1 Introduction

It is hypothesized that a prodrug at the N-terminal end of GLP will extend and improve the pharmacodynamics of this peptide hormone (Fig. 1.5a and b).

The ideal prodrug should be soluble in water at a pH of 7.2 and 37 °C, and it should be stable in the powder form for long-term storage. It should be immunologically

A. De, *Application of Peptide-Based Prodrug Chemistry in Drug Development*, 15
SpringerBriefs in Pharmaceutical Science & Drug Development,
DOI: 10.1007/978-1-4614-4875-4_2, © The Author(s) 2013

silent and biologically inactive when injected in the body, and be quantitatively converted to the active drug within a defined period of time. Our greatest interest lies in prodrugs with a $t_{1/2}$ of between 10 and 100 h (weekly, or even monthly duration) under physiological conditions (pH of 7.2 and 37 °C).

During the course of this research, we identified four GLP analogs that are physically and chemically stable, and whose conversion from prodrug to the active drug form under normal physiological conditions ranges within the optimal range. One of these peptides converts to GLP with a half-life of 64 h. The in vivo extension in duration of action of this magnitude would constitute the longest acting peptide prodrug ever designed. These analogs possess a minimal alteration to the native amino acid sequence, and this should minimize potential adverse immunogenic affects. The bioactivities of these synthetic peptides have been determined using in vitro cellular assays.

At the beginning of this research, many strategies were considered for constructing our prodrugs. It was contemplated that the protecting groups of the prodrug could be cleaved by enzymes as reported for nucleotide prodrugs [1]. However, we decided to design a prodrug that would slowly convert to the parent drug at physiological conditions of 37 °C and pH 7.2 driven by inherent chemical instability. The pH and temperature were relied upon for this conversion, as they are virtually and physiologically invariant. We are seeking a prodrug that converts quantitatively to the drug under physiological conditions without the aid of any enzyme. The establishment of prodrug chemistry at the N-terminal end of GLP should be translatable for use with other peptides where this specific site is vital to bioactivity.

We decided to synthesize a chemical derivative of GLP that would convert to GLP spontaneously as stated above. A few possible choices were contemplated before deciding that our primary target in prodrug chemistry would be diketopiperazine formation. The N-terminal histidine is important for the potency of our drug, and a reversibly modified histidine side chain was considered, as has been reported for thyrotropin-releasing hormone (TRH) [2]. However, we dispensed with this approach since this prodrug chemistry is specific to the imidazole ring. This means that the prodrug chemistry requires the presence of a histidine, and possibly also at an N-terminal site.

Thiol esters could also be synthesized by thioesterification [3]. However, a thiol ester is relatively unstable in physiology and thus would likely cleave too fast, thereby resulting in rapid clearance of the prodrug in its active form. Additionally, the use of thiol-based chemistry is fraught with other difficulties, such as instability of disulfide bonds. We decided to forgo this course of action.

Our work therefore focused on making amide and ester prodrugs at the N-terminus that upon cleavage of the suitable amide or ester bond generates the desired drug. Esters are normally more labile than amides; however they are easily hydrolyzed by the ubiquitous serum esterases [4]. Hence, amide bond based prodrugs were a much more attractive design, although the risk of peptidase degradation may potentially complicate in vivo application.

In considering amide prodrugs, it is reported that histidyl-proline amide cyclizes to a cyclo (His-Pro) [5] at a pH of 7 and 37 °C with a $t_{1/2}$ of 140 min, with the release of NH_3. The imidazole ring is purported to be playing a catalytic role at this pH. Hence, other dipeptides that did not have a histidine cleaved more slowly, and tripeptides did not cleave at all [5]. Thus, it seemed from this work that there might be a basic difference in the rate of diketopiperazine formation from the cleavage of a secondary amide (as in a tripeptide) as compared to a primary one (as in a dipeptide amide).

In another paper [6], the intramolecular aminolysis of Phe-Pro-p-nitroanilide (Phe-Pro-pNA) to Phe-Pro-diketopiperazine (Phe-Pro-DKP) was studied as a function of pH. The pH-rate plot showed that the rate of the formation of the DKP was dependent on the degree of ionization of the N-terminal amino group, with the unprotonated free amine being more reactive than the protonated form. In their experiment, the authors used an activated, strongly electron withdrawing p-nitroanilide dipeptide instead of a natural tripeptide. This was because these p-nitroanilide dipeptides dissociate by DKP formation more rapidly than amino acid amides (i.e., natural dipeptides), thereby greatly facilitating their use in kinetic studies. The calculated $t_{1/2}$ of conversion of a Gly-Pro-p-nitroanilide (Gly-Pro-pNA) dipeptide to Gly-Pro-diketopiperazine (Gly-Pro-DKP) under physiological conditions was about 120 h. This is consistent with the previous assertion that there is a significant difference in the dissociation rate between a primary (around 140 min as in previous example [5]) and an activated secondary amide. It is also important to note that in a natural tripeptide, the half-life of the DKP formation would be further extended since there is no electronic assistance from the pNA.

In both these papers [5, 6], it seems that the presence of proline in the C terminus of the dipeptide extension accentuates the formation rate of the DKP. This is likely due to contribution of the cis-proline conformer in the facilitation of the dipeptide's adoption of an optimal steric conformation for formation of DKP. The observations with modifications of an N-terminal residue upon the rate of DKP formation at pH 7.0 have been more varied. Reports suggest that it might depend on the pKa of the residue [5], on its bulk[7], or on the conformational stability of the resulting DKPs [6].

We also envisioned modification of a hydroxyl group at what otherwise would be the N-terminus to prepare a depsi-peptide and thus make an ester prodrug. An ester can cleave hydrolytically [8, 9] or via the formation of five- [17, 18] or six-membered rings (like DKP). The ester prodrugs of floxuridine (FUdR) [9] convert by general ester hydrolysis, i.e., a nucleophilic attack by water on the ester carbonyl. For the most part, the bulk of the pro-moiety influenced the hydrolysis of the FUdR prodrug (the Val ester prodrugs dissociated the slowest). In another paper, the authors studied the cyclization of the dipeptide esters in paracetamol to form a diketopiperazine [10]. They observed that they could obtain differential time action depending on the dipeptide structure. They also considered the possibility that in paracetamol, the drug release might have been via the general mechanism of ester hydrolysis and not the formation of a DKP ring. However, they eliminated it as in that case the nature of the dipeptide would have a lesser effect

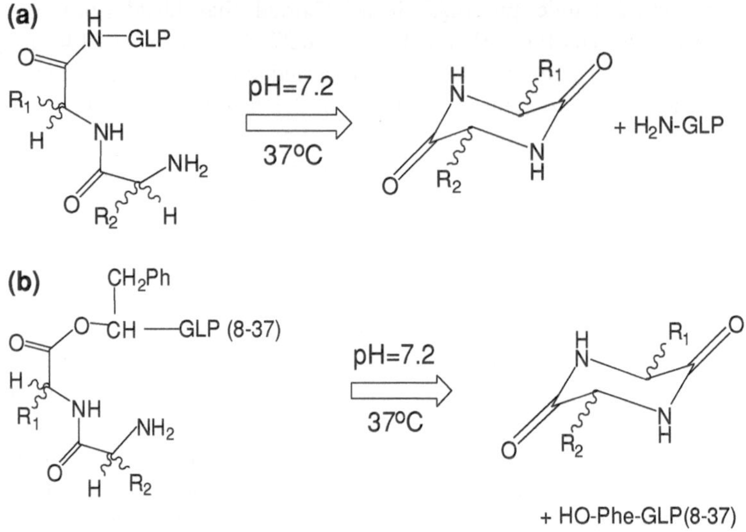

Fig. 2.1 Cleavage of amide (**a**) and ester prodrugs (**b**)

on reactivity. The proof of labile esters cleaving very fast under physiological conditions is exemplified by the fast cleavage of the dipeptide esters in paracetamol where all the prodrugs had a $t_{1/2}$ of less than 20 min [10].

In another investigational report, this time with cyclosporine-A prodrugs [11, 12] it has also been seen that through modulating the chemical nature of dipeptide esters it was possible to get conversion rates at physiological conditions ranging from minutes to several hours, but not longer. In these papers [10, 11], it seems that the presence of a minimally bulky glycine residue in the C terminus of the dipeptide extension accentuates the rate of formation of the DKP. This might be because of less bulk and the preferred conformational effect of glycine. However in the cyclosporine prodrug [11], it seems unexpected that the presence of a proline in the C terminus of the DKP actually attenuates the rate of the conversion.

In another investigation [13], the C terminal amides of glycine were rapidly hydrolyzed at 25 °C and a pH of 7 when the N-terminus was N-hydroxyethylated. The $t_{1/2}$ of bis-N-2-hydroxyethylglycinamide is three hours. In this case, the C terminal amide bond is activated by H bonding with the N-hydroxyethyl group. However, there was no practical way that one could modify the structure of the N-hydroxyethyl group as the precise Vander Waal radii were required to activate the amide group.

Consequently, we explored the intramolecular cyclization reaction of dipeptide esters and amides to form diketopiperazine as an example of a chemoreversible prodrug (Fig. 2.1). This chemistry is reasonably straightforward and allows at least four points (stereochemistry of $R1$ and $R2$, nature of the nucleophile and leaving group as shown in Fig. 2.1a and b) where structure can be stereochemically controlled to refine the rate of formation with release of the active peptide.

Finally, they can also be prepared from readily available alpha-amino acids using established chemistry [14].

As shown below (Fig. 2.1a and b), prodrugs of varying half-lives were designed by modifying $R1$ and $R2$. In the reaction below, there is an N-terminal histidine residue (native peptide), and an amide bond is being broken (Fig. 2.1a). In the second equation, the ester bond of phenyllactic acid (hydroxyl phenylalanine) is being broken (Fig. 2.1b). Though it might be beneficial to use the histidine in the 7th position of GLP, alternatives to the native N-terminal histidine were utilized for synthetic and analytical ease. Once the chemistry of a longer acting prodrug is established, it is plausible to return to the histidine or for that matter any other suitably potent amino acid at the N-terminus.

Thus, we propose to make a prodrug that will slowly convert ($t_{1/2}$ between 10 and 100 h) to the parent drug at physiological conditions of 37 °C and pH 7.2 so that they could be administered at a weekly or monthly frequency. As far as possible, a native sequence shall be used in the prodrug so as to minimize the chances of an immunological response. We rely on the pH and the temperature for this intramolecular conversion, as they are virtually invariant. It is of essence to note that this reaction should be concentration independent.

Four such peptides were identified with protracted half-lives, with minimal potency as compared to the drug in the luciferase-based bioassay. These prodrugs regained their potencies after incubation in PBS buffer at a pH of 7.2 and temperature of 37 °C.

Prodrugs in this book can be broadly classified into four different types:

1. An amine nucleophile cleaving an amide bond (Class 1): This will dissociate with the formation of the corresponding 2,5-diketopiperazine.

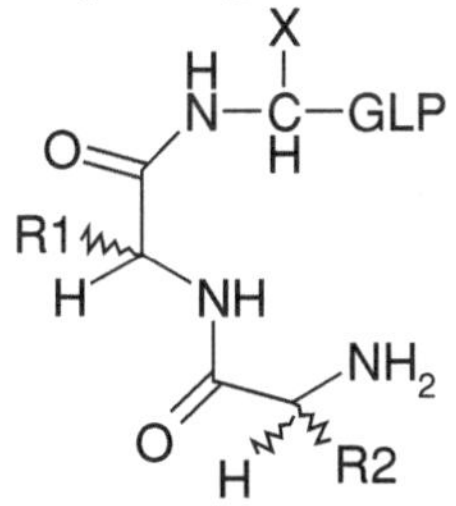

2. A hydroxyl nucleophile cleaving an amide bond (Class 2): This will dissociate with the formation of the corresponding 2,5-diketomorpholine.

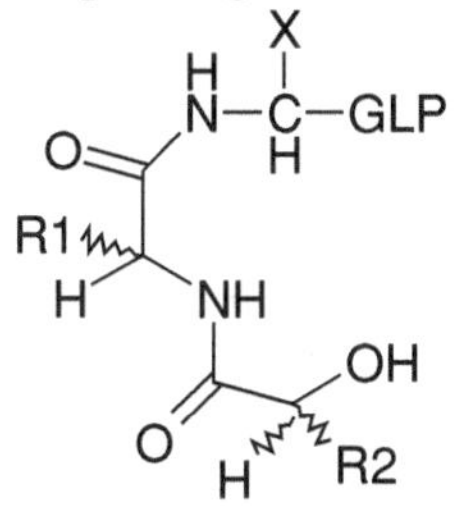

3. An amine nucleophile cleaving an ester bond (Class 3): This will dissociate with the formation of the corresponding 2,5-diketopiperazine.

4. A hydroxyl nucleophile cleaving an ester bond (Class 4): This will dissociate with the formation of the corresponding 2,5-diketomorpholine.

The initial analysis was performed with the crude peptides. The rate of this reaction was not different from the pure peptides as this is an intramolecular cyclization. Additionally, it is also possible to observe our molecules of interest (both the prodrug and the drug) with a HPLC and MALDI analysis even in the midst of contaminating material. In most cases, an excellent mass balance for the disappearance of the prodrug and the appearance of the drug was observed. After we were satisfied that the crude prodrug had the required $t_{1/2}$, the prodrug was purified and a standard luciferase-based bioassay was conducted to obtain relative potencies.

The experimental design of our bioassay was based on the general principles of "reporter gene technology" [15] (Fig. 2.2). In our case, the luciferase-based reporter gene assay for cAMP detection was used. The changes in the intracellular cAMP concentrations [16] caused by the GLP receptor-mediated interactions are detected by the changes in the expression level of the luciferase gene. The transcription of this gene is regulated by the cAMP response-element binding protein (CREB) binding to cAMP response element (CRE).

This is an artificially created test system where the luciferase gene is downstream to the CRE which resembles nature's response system all the way up to the point of gene expression where the luciferase gene is expressed. This modification is necessary as the concentration of activated luciferase is easier to measure than that of cAMP.

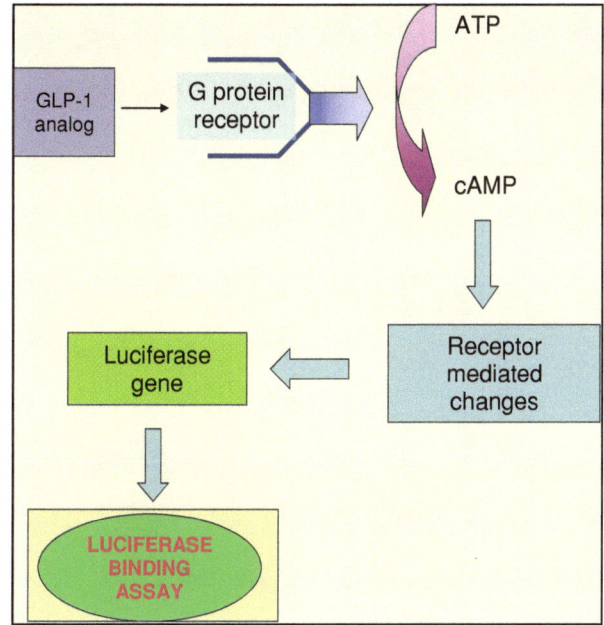

Fig. 2.2 Pictorial explanation of luciferase-based reporter gene assay

References

1. Poijarvi-Virta P (2006) Curr Med Chem 13(28):3441–3465
2. Bundgard H, Moss J (1990) Pharm Res 7(9):885–892
3. Oliyai R, Valentino JS (1993) Annu Rev Pharmacol Toxicol 33:521–544
4. Stañczak A, Ferra A (2006) Pharmacol Rep 58:599–613
5. Bundgard H, Moss J (1990) J Pharm Pharmacol 42:7–12
6. Goolcharran C, Borchardt RT (1998) J Pharm Sci 87(3):283–288
7. Borchardt RT, Cohen LA (1972) J Am Chem Soc 94:9166–9174
8. Larsen SW, Anderson M, Larsen C (2004) Eur J Pharm Sci 22(5):399–408
9. Balvinder S et al (2003) Pharm Res 20(9):1381–1388
10. Santos C et al (2005) Bioorg Med Chem Lett 15:1595–1598
11. Hamel AR et al (2004) J Pept Res 63:147–154
12. Hamel AR, Hubler F, Mutter M (2005) J Pept Res 65:364–374
13. Suggs JW, Pires RM (1997) Tetrahedron Lett 38:2227–2230
14. Fischer PM (2003) J Pept Sci 9(1):9–35
15. Louise HN (1999) Biochem Pharmacol 58:749–757
16. Williams C (2004) Nature reviews. Drug Discov 3:125–135
17. Hamada Y et al (2004) Bioorg Med Chem 12:159–170
18. Hamada Y et al (2002) Bioorg Med Chem 10:4155–4167

Chapter 3
Experimental Procedures

Abstract Various procedures for synthesis of peptides by Boc and Fmoc-based methods are described. Synthesis of depsipeptides and N-terminal hydroxyl peptide synthesis are illustrated. The various peptides are characterised by mass spectrometry and analytical HPLC. The rate constant for the dissociation of the respective prodrugs can be studied by chromatographic procedures. The concentrations of the prodrug and the drug can be estimated from their peak areas respectively. The first-order dissociation rate constants of the prodrugs can be determined by plotting the logarithm of the concentration of the prodrug at various time intervals. Peptides can be purified using preparative HPLC and purified peptides can be used for luciferase–based bioassays to study the activity of the various prodrugs. Detailed bioassays to examine the conversion of the prodrug to the drug under therapeutic conditions are described.

Keywords Boc · Fmoc · Depsipeptide · N-terminal hydroxyl peptide · MALDI · Mass spectrometry · High Pressure Liquid Chromatography · HPLC · Peptide bioassay · Luciferase

The standard procedure is described briefly here, and the details are discussed later. PAM resin (PAM resin is OCHE2-phenylacetamidomethyl–copolystyrene-1 % divinylbenzene), (100–180 mesh, 1 % DVB cross-linked polystyrene; loading of 0.7–1.0 mmol/g), Boc-protected and Fmoc protected amino acids were purchased from Midwest Biotech. Other reagents such as the α-hydroxy acids (phenyllactic acid and glycolic acid) were purchased from Aldrich. The solid phase peptide syntheses using Boc-protected amino acids were performed on an Applied biosystem 430A peptide synthesizer [1]. Fmoc protected amino acid synthesis was performed using the Applied biosystems model 433 peptide synthesizer. The manual synthesis of depsipeptides was performed in sintered reaction vessels using analogous procedures [1, 2].

A. De, *Application of Peptide-Based Prodrug Chemistry in Drug Development*, 23
SpringerBriefs in Pharmaceutical Science & Drug Development,
DOI: 10.1007/978-1-4614-4875-4_3, © The Author(s) 2013

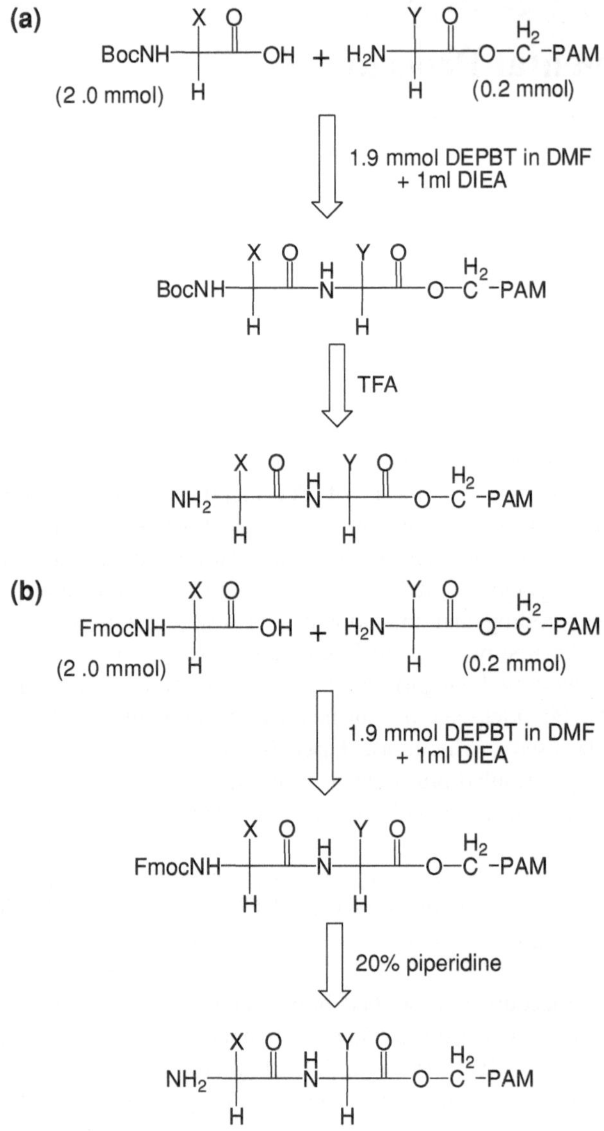

Fig. 3.1 a Peptide synthesis (Boc method). **b** Peptide synthesis (Fmoc method)

3.1 Peptide Synthesis (Boc Amino Acids/HF Cleavage)

The bulk of the synthesis was accomplished by using this method (Fig. 3.1a).

Synthesis of these analogous was performed on the Applied biosystem model 430A peptide synthesizer. Synthetic peptides were constructed by sequential addition of amino acids, and activated esters of each amino acid were generated by the addition of 1.9 mmol (3.8 ml of a 0.5 M solution) of 3-(Diethoxy-phospho-ryloxy)-3H-benzo[d][1,2,3] triazin-4-one (DEPBT) in DMF to a cartridge containing 2 mmol of Boc protected amino acid. The amino acids were dissolved by bubbling nitrogen gas through the cartridge. 1 ml of N,N- Diisopropylethyl-amine was added to the cartridge to effect ester formation. This solution was transferred to the reaction vessel containing the 0.2 mmol of the C-terminal residue attached to the PAM resin, vortexed several times, and allowed to couple to the resin for 10 min. After washing to remove the unreacted reagents, the N-terminal Boc protecting group was removed by treatment with trifluoroacetic acid (TFA) for 5 min. The resin was washed with DMF and the cycle was repeated for the desired number of steps until the chain was assembled. The reaction vessel at the end of the synthesis (typically 30 amino acids) contained approximately 1.2–1.5 g of protected peptidyl-PAM resin. The resin was washed numerous times with dimethylformamide (DMF), treated with trifluoroacetic acid to remove the last t-Boc protecting group and finally washed several additional times with DMF, dichloromethane (DCM) and dried.

The peptidyl-resin was treated with anhydrous HF (procedure explained later in this section), and this typically yielded approximately 350 mg ($\sim$50 % yield) of a crude deprotected-peptide.

3.2 Peptide Synthesis (Fmoc Amino Acids/HF Cleavage)

This synthesis scheme was performed manually with a few amino acids at selective sites. Here, it is to be noted that although FMOC chemistry has been used in the synthesis, the peptides have always been built on PAM resin that required treatment with HF to cleave the peptide from the solid support. The yield of these peptides is approximately as stated earlier for Boc/PAM synthesis.

The synthesis was carried out as described in the previous section. At the end of the coupling step, the peptidyl-resin was treated with 20 % piperidine to remove the N-terminal Fmoc protecting group. It was washed repeatedly with DMF and this repetitive cycle was repeated for the desired number of coupling steps. The peptidyl-resin at the end of the entire synthesis was dried by using DCM, and the peptide was cleaved from the resin with anhydrous HF.

3.3 Depsipeptide Synthesis (Amino Ester Formation)

These syntheses[2] were performed manually (Fig. 3.2). In this case, the peptidyl-resin had an α-hydroxyl-N terminal extension instead of a N-terminal amine and the acylation was done at the α-hydroxyl group. This reaction takes a longer time than that of the amide bond formation, as the hydroxyl group is a weaker nucle-ophile as compared to the amine. The reaction time was typically 12 h.

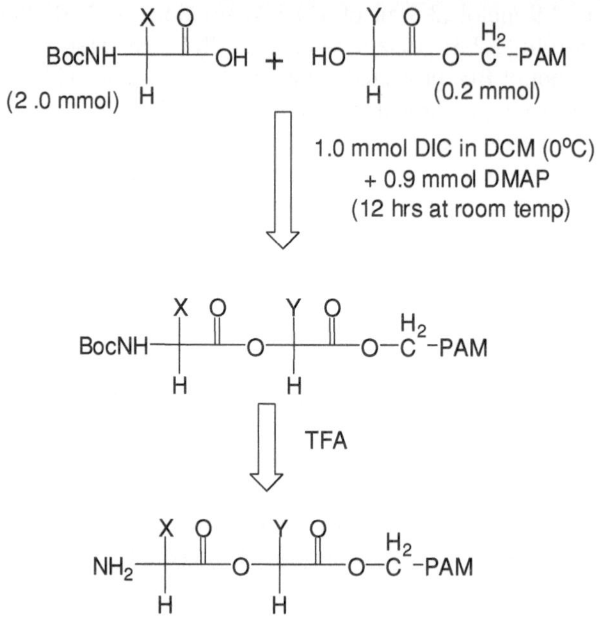

Fig. 3.2 Depsipeptide synthesis (Amino ester formation)

Initially, the activated esters of each amino acid were generated by the addition of 1 mmol (0.155 ml of Diisopropylcarbodiimide (DIC) to a cartridge containing a solution of 2 mmol of Boc protected amino acid residue in 2 ml DCM. This was cooled to 10 °C for 10 min. 0.9 mmol (244 mg) of dimethylaminopyridine (DMAP) was added to the cartridge to accelerate ester formation. This mixture was transferred to the reaction vessel containing the peptidyl-resin upon which the peptide was synthesized. The reaction vessel was stirred for 12 h.

The peptidyl-resin was dried using DCM and the synthesis of the desired peptide was continued. The peptidyl-resin at the end of the entire synthesis was dried by using DCM, and finally treated with anhydrous HF to generate the desired peptide.

3.4 N-Terminal Hydroxyl Peptide Synthesis (α-Hydroxyl-N Terminal Extension)

In this reaction, the free amine of the peptidyl-resin reacts with an α-hydroxyl acid to form an α-hydroxyl-N terminal extension [3] (Fig. 3.3).

In this regard, only two such α-hydroxyl acids were used namely, glycolic acid (OH-glycine) and phenyllactic acid (OH-phenylalanine). These syntheses were also performed manually. The peptides were constructed by addition of the

Fig. 3.3 α hydroxyl-N
terminal extension

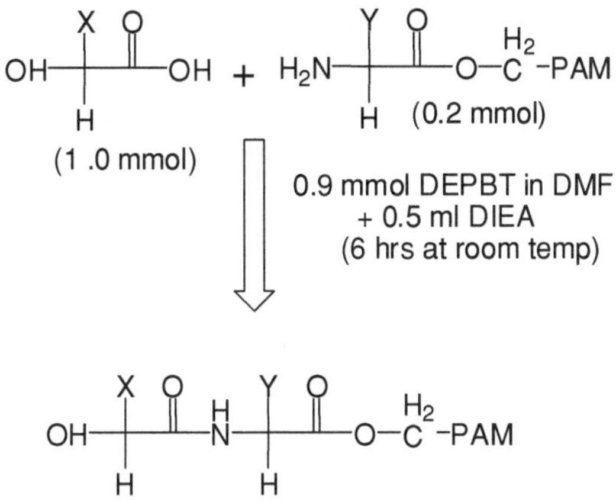

α-hydroxyl acid, and activated esters of the α-hydroxyl acid were generated by the addition of 0.9 mmol of DEPBT (270 mg) to a cartridge containing a solution of 1 mmol of Boc protected residue in 2 ml DMF. 0.5 ml of DIEA (N, N-Diisopropylethylamine) was added to the cartridge to accelerate ester formation. This mixture was transferred to the reaction vessel containing the peptidyl-resin upon which the peptide was synthesized. The reaction time was 6 h.

The peptidyl-resin was dried using DCM and the synthesis of the desired peptide was continued. The peptidyl-resin at the end of the entire synthesis was dried by using DCM, and cleaved by anhydrous HF to generate the free peptide.

3.5 HF Treatment of the Peptidyl-Resin

The peptidyl-resin (30–200 mg) was placed in the hydrogen fluoride (HF) reaction vessel for cleavage. 500 μL of p-cresol was added to the vessel as a carbonium ion scavenger. The vessel was attached to the HF system and submerged in the methanol/dry ice mixture. The vessel was evacuated with a vacuum pump and 10 ml of HF was distilled to the reaction vessel. This reaction mixture of the peptidyl-resin and the HF was stirred for one hour at 0 °C, after which a vacuum was established and the HF was quickly evacuated (10–15 min). The vessel was removed carefully and filled with approximately 35 ml of ether to precipitate the peptide and to extract the p-cresol and small molecule organic protecting groups resulting from HF treatment. This mixture was filtered utilising a teflon filter and repeated twice to remove all excess cresol. This filtrate was discarded. The precipitated peptide dissolves in approximately 20 ml of 10 % acetic acid (aq). This filtrate, which contained the desired peptide, was collected and lyophilised.

3.6 Analysis Using Mass Spectrometry

The mass spectra were obtained using a Sciex API-III electrospray quadrapole mass spectrometer with a standard ESI ion source. Ionisation conditions that were used are as follows: ESI in the positive-ion mode; ion spray voltage, 3.9 kV; orifice potential, 60 V. The nebulising and curtain gas used was nitrogen flow rate of 0.9L/min. Mass spectra was recorded from 600–1800 Thompsons at 0.5 Th per step and 2 ms dwell time. The sample (about 1 mg/mL) was dissolved in 50 % aqueous acetonitrile with 1 % acetic acid and introduced by an external syringe pump at the rate of 5 μL/min.

When the peptides were analyzed in PBS solution by ESI MS, they were first desalted using a ZipTip solid phase extraction tip containing 0.6 μL C4 resin, according to instructions provided by the manufacturer. (http://www.millipore.com/catalogue.nsf/docs/C5737).

3.7 High Pressure Liquid Chromatography (HPLC) Analysis

Preliminary analyses were performed with these crude peptides to get an approximation of their relative conversion rates in phosphate buffered saline (PBS) buffer (pH, 7.2) using high performance liquid chromatography (HPLC) and MALDI analysis. The crude peptide samples were dissolved in the PBS buffer at a concentration of 1 mg/ml. 1 ml of the resulting solution was stored in a 1.5 ml HPLC vial which was then sealed and incubated at 37 °C. Aliquots of 100 μl were drawn out at various time intervals, cooled to room temperature and analyzed by HPLC.

The HPLC analyses were performed using a Beckman system gold chromatography system using a UV detector at 214 nm. HPLC analyses were performed on a 150 × 4.6 mm C18 Vydac column. The flow rate was 1 ml/min. Solvent A contained 0.1 % TFA in distilled water, and solvent B contained 0.1 % TFA in 90 % CH$_3$CN. A linear gradient was employed (40–70 % B in 15 min). The data was collected and analyzed using peak simple chromatography software.

The initial rates of hydrolysis were used to measure the rate constant for the dissociation of the respective prodrugs. The concentrations of the prodrug and the drug were estimated from their peak areas respectively. The first-order dissociation rate constants of the prodrugs were determined by plotting the logarithm of the concentration of the prodrug at various time intervals. The slope of this plot gives the rate constant 'k'. The half lives of the degradation of the various prodrugs were then calculated by using the formula $t_{1/2} = 0.693/k$.

3.8 Preparative Purification Using HPLC

After we were satisfied that the prodrug had an appropriate $t_{1/2}$, the prodrug was purified. The purification was performed using HPLC analysis on a silica-based 1×25 cm Vydac C18 (5μ particle size, 300A° pore size) column. The instruments used were: waters associates model 600 pump, injector model 717 and UV detector model 486. A wavelength of 214 nm was used for all samples. Solvent A contained 10 % CH_3CN/0.1 % TFA in distilled water, and solvent B contained 0.1 % TFA in CH_3CN. A linear gradient was employed (0–100 % B in 2 h). The flow rate is 1.2 ml/min and the fraction size was 6 ml. From ~ 350 mgs of crude peptide, 80 mgs of the pure peptide (~ 23 % yield) was typically obtained.

3.9 Bioassay Experimental Design: Luciferase-Based Reporter Gene Assay for cAMP Detection

The ability of each GLP analog or prodrug to induce cAMP was measured [4] in a firefly luciferase-based reporter assay (Fig. 2.2). The cAMP production that is induced is directly proportional to the GLP binding to its receptor. HEK293 cells co-transfected with the GLP receptor and luciferase gene linked to cAMP responsive element were employed for bioassay.

The cells were serum-deprived by culturing 16 h in Dulbecco minimum essential medium (invitrogen, Carlsbad, CA) supplemented with 0.25 % bovine growth serum (HyClone, Logan, UT) and then incubated with serial dilutions of either GLP analogs or prodrugs for 5 h at 37 °C, 5 % CO_2 in 96 well poly-D-Lysine-coated "Biocoat" plates (BD Biosciences, San Jose, CA). At the end of the incubation, 100 μL of luclite luminescence substrate reagent (Perkin Elmer, Wellesley, MA) were added to each well. The plate was shaken briefly, incubated 10 min in the dark and light output was measured on MicroBeta-1450 liquid scintillation counter (Perkin-Elmer, Wellesley, MA). The effective 50 % concentrations (EC_{50}) were calculated by using origin software (OriginLab, Northampton, MA).

References

1. Schnolzer M et al (1992) Int J Pept Protein Res 40(3–4):180–193
2. Personal Communications with Prof. Phil Dawson . Scripps Research Institute, San Diego, California
3. Williams C (2004) Nature Reviews. Drug Discovery 3:125–135
4. Samantha EG, Peter JB, Louise HN (1997) J Biomol Screen 2(4):235–240

Chapter 4
Characterization of Prodrugs

Abstract Synthesis of a model peptide GLP1-oxyntomodulin is described to demonstrate that relatively large peptides can be synthesized using methods in the last chapter. The synthesis of Classes 1–4 prodrugs using various drug scaffolds are described. The structural nature of the side chain (especially β branching as evidenced in dipeptides containing valine and *tert*-butyl glycine); the stereochemistry, and the pKa of the nucleophile serve an important role in determining the relative rate of cleavage. Several fast and slow acting ester prodrug candidates were tested for their potency and were found to have minimal potency as compared to the drug. The prodrugs regained their potencies after incubation in PBS buffer at a pH of 7.2 and temperature of 37 °C.

Keywords Oxyntomodulin · GLP-1 · Prodrug · Luciferase · Nucleophile · DKP · DMP · Active imidazole · Pharmacodynamics · Therapeutic index

4.1 GLP-Oxyntomodulin

The GLP-oxyntomodulin chimeric peptide shown below was synthesized. The last eight amino acids were derived from oxyntomodulin and are shown in italic.

HAEGTFTSDVSSYLEGQAAKEFIAWLVKGRG*KRNRNNIA*

The rationale behind synthesizing this chimeric peptide is two fold. First, to demonstrate that a peptide of a mass of around 4,300 Da could be synthesized. Second, if this peptide is biologically potent, then a GLP analog with an extended C-terminus could be used for any future modification that might be required. This peptide has a mass of 4,322.5 Da (Fig. 4.1a). In this way, the synthesis of GLP-oxyntomodulin was confirmed. The receptor binding activity of GLP-oxynto-modulin was determined in the GLP-receptor Luciferase assay (Fig. 4.1b).

A. De, *Application of Peptide-Based Prodrug Chemistry in Drug Development*,   31
SpringerBriefs in Pharmaceutical Science & Drug Development,
DOI: 10.1007/978-1-4614-4875-4_4, © The Author(s) 2013

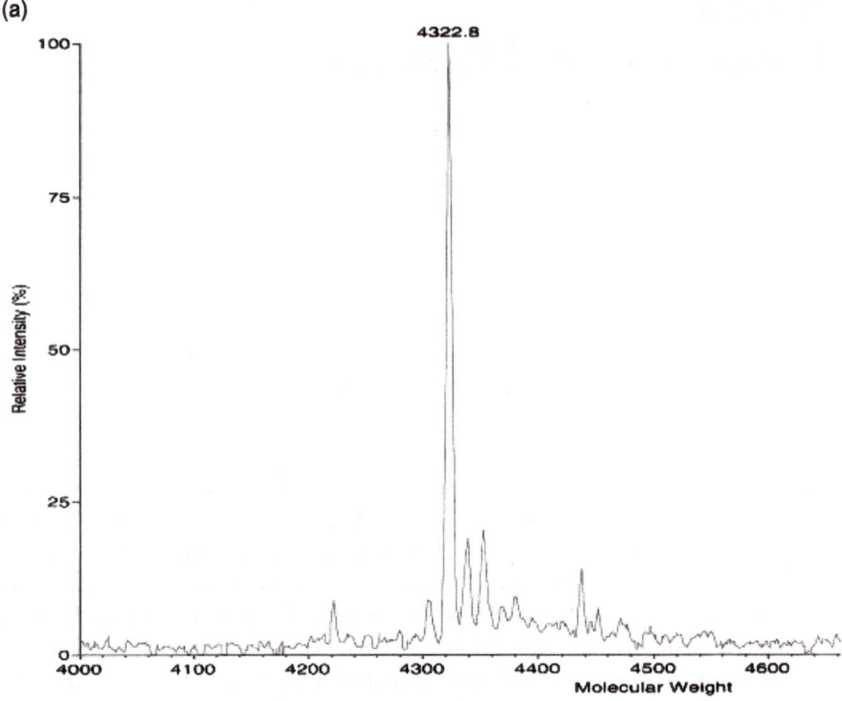

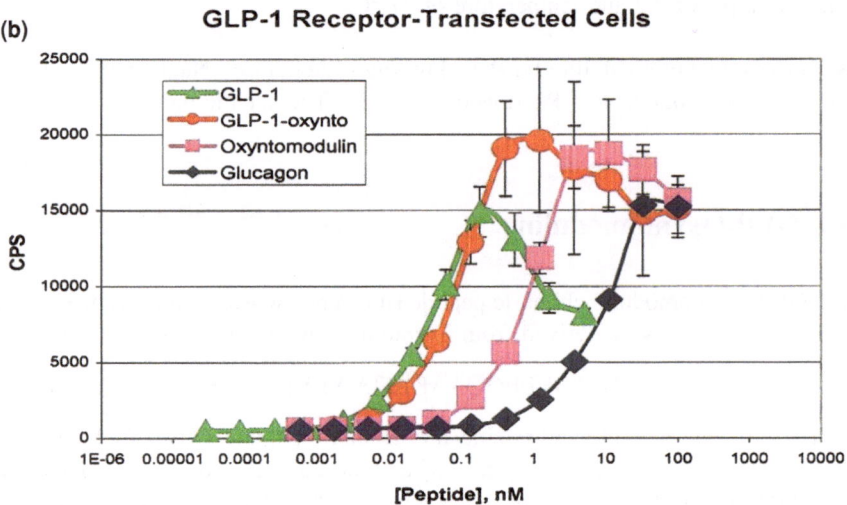

Fig. 4.1 a Mass spectra of purified GLP-oxyntomodulin chimeric peptide. **b** Bioassay results of GLP-oxyntomodulin chimeric peptide

The GLP-oxyntomodulin was found to be at least as potent as the native GLP peptide in the luciferase assay. The GLP-oxyntomodulin also has a higher apparent maximal efficacy as compared to GLP (Fig. 4.1b). This observation warrants additional study as the less potent oxyntomodulin also demonstrated super-efficacy. This signifies that the first portion of the project is successful and we focused on the central element of this study, specifically the N-terminal prodrug design.

4.2 Adding Dipeptides to the N-Terminus

Dipeptides were covalently attached to the N-terminus of GLP (sequence in Fig. 1.3b) to study differential tendencies for intramolecular cyclization and cleavage through diketopiperazine formation. This is represented schematically in Fig. 4.2a. All Class I prodrugs were synthesized by analogous procedures.

The biologically inactive dipeptide extended GLP was converted to the active GLP upon cleavage of the amide bond along with the DKP (Fig. 4.2b; Table 4.1). In Fig. 4.2c (Tables 4.2, 4.3), the same conversion is shown with the phenylalanine in the 1st position of GLP. Prodrugs of varying half lives were envisioned by chemically modifying the R_1 and R_2 positions. This validates our reason for probing the DKP formation strategy.

The results of the cleavage in the prodrugs are shown in Table 4.1. The first peptide synthesized was named $G^5P^6H^7$, GLP(8-37) where the R_1 is the side chain for proline and R_2 refers to the side chain for glycine (peptide 1 in Table 4.1). All peptides mentioned, hereafter, will have the same systematic nomenclature and the stereochemistry was assumed to be the L-isomer unless otherwise stated. The peptide was prepared synthetically by solid phase synthesis as described earlier. The synthesis was confirmed by MALDI-MS analysis (3509.5 Da) as shown in Fig. 4.3.

To explore the possible formation of DKP, and simultaneous regeneration of the H^7, GLP(8-37), the $G^5P^6H^7$, GLP(8-37) was incubated in PBS buffer at 37 °C for approximately a week. Additionally, the peptide was heated at 100 °C to accelerate the amide bond cleavage. Analysis by reverse phase HPLC showed no apparent cleavage of the amide bond in this first set (Table 4.1). The structures of the peptides in Table 4.1 are illustrated in Appendix II.

To investigate the propensity of various prodrugs to undergo DPK formation, all subsequent peptides were similarly synthesized and analyzed by MS and HPLC following treatment at 100 and 37 °C as described above.

The dipeptide extension of peptides 1 and 2 in Table 4.1 were synthesized to facilitate DPK formation by sterically assisting in the cleavage of the amide bond. It was thought that the *cis*-orientation of proline would contribute in the facilitation of the dipeptide's adoption of an optimal steric conformation for the formation of DKP. However, the amide bond is quite robust and did not cleave. In peptides 3–6, an acid–base catalyzed general hydrolysis of the amide bond was attempted. Since, the leaving group will be the histidine at the N-terminus, it was purported that the

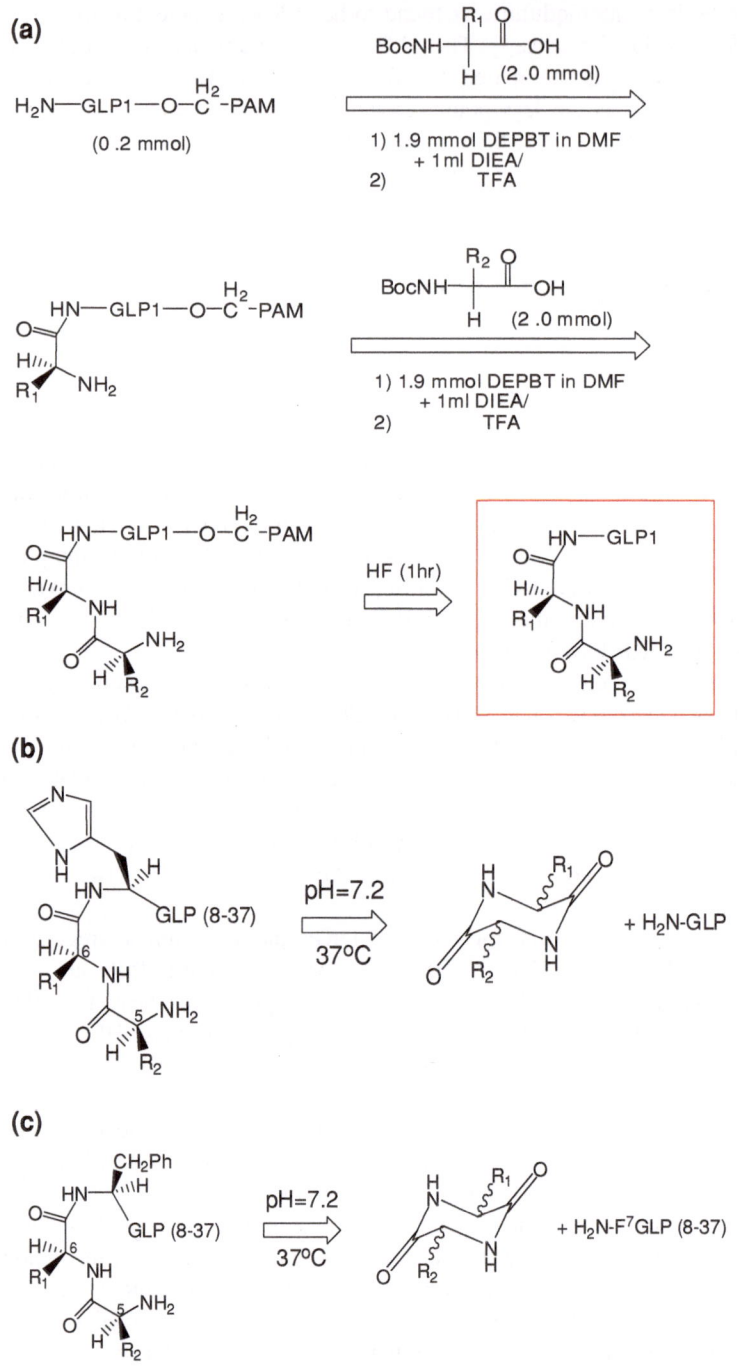

Fig. 4.2 a Schematic synthesis of Class I prodrugs. **b** Cleavage of amide bond to form DKP + H[7], GLP(8-37). **c** Cleavage of amide bond to form DKP + F[7], GLP(8-37)

Table 4.1 Attempted cleavage of H^7, GLP(8-37) prodrugs to form native GLP-1

Serial no	Peptide (Xaa^{1}Yaa^2-GLP1)	Rate of cleavage at 100 °C	Rate of cleavage at 37 °C
1	GP	No cleavage	No cleavage
2	PP	No cleavage	No cleavage
3	γ Glu	No cleavage	No cleavage
4	E	No cleavage	No cleavage
5	P	No cleavage	No cleavage
6	H	No cleavage	No cleavage
7	PH	No cleavage	No cleavage

In 4–6, there is just a single amino acid added to the N-terminus of GLP-1. In 7, the dipeptide modification was made to sandwich the carbonyl bond of interest between two histidines. In all these cases, there is histidine in the first position (histidyl leaving group) of GLP-1

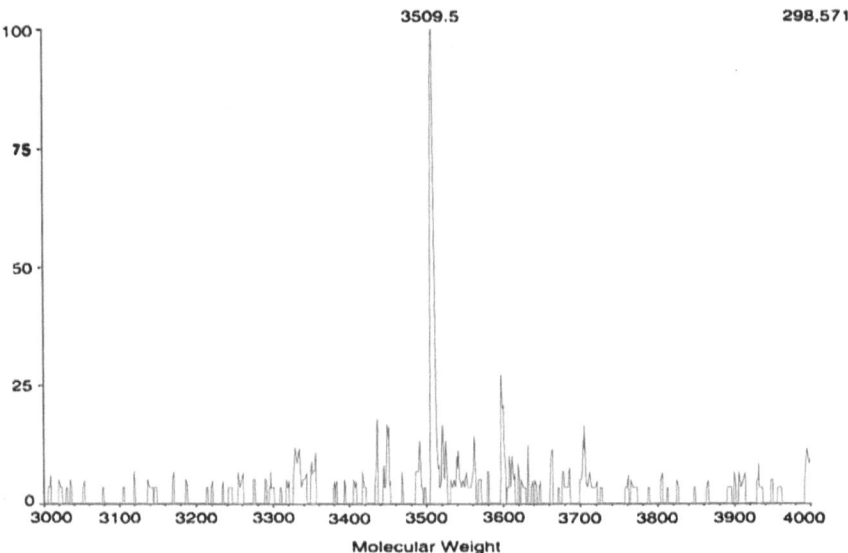

Fig. 4.3 Mass spectra of $G^5P^6H^7$, GLP(8-37)

imidazole ring might in some way assist in the cleavage by general acid–base catalysis. In 7, the carbonyl bond of interest is sandwiched between two histidines (amino acid Y and the histidyl leaving group). This design was directed at a proton assisted cleavage of the amide bond via diketopiperazine formation. But even this prodrug did not cleave.

At this point, it was speculated that perhaps the imidazole nucleus was playing an attenuating role in the cleavage of the amide bond. To test this possibility, a different leaving group was studied. F^7, GLP(8-37) was synthesized and purified using the standard procedure described above. It was determined to be a full agonist with 10 % the potency of native GLP. Dipeptides were added to this GLP analog to study the same type of reaction as described above (Table 4.2; reaction

Table 4.2 Attempted cleavage of dipeptide extended F^7, GLP(8-37)

Serial no	Peptide (Xaa^{1}Yaa^2-F^{7a}-GLP8-37)	Rate of cleavage at 100 °C	Rate of cleavage at 37 °C
1	GP	No cleavage	No cleavage
2	GSar^b	No cleavage	No cleavage

^{a} In all these cases, there is phenylalanine in the first position (phenylalanyl leaving group) of GLP
^{b} Sar represents sarcosine

shown in Fig. 4.2c). The structures of the peptides in Table 4.2 are drawn out in Appendix III.

The dipeptide extension of peptides 1 and 2 (Table 4.2) were designed to sterically assist in the cleavage of the amide bond. In peptide 2, sarcosine was used [1], as it has been previously reported to enhance the rate of cleavage. However, the amide bond remained resistant to cleavage.

A minor difficulty was encountered at this point. It seemed that the F^7, GLP(8-37) analogs were not very soluble in PBS at 37 °C. Hence, we focused on a modified GLP analog; GLP(7-36)-CEX amide where the C-terminus is a serine amide. The CEX sequence is the C-terminal nine amino acids of exendin-4. The last nine amino acids were derived from exendin-4 and are shown in italic.

$$H^7AEGTFTSDVSSYLEGQAAKEFIAWLVKGR\textit{PSSGAPPPS}\text{-}\textit{amide}$$

This peptide been observed in our laboratory to be ten times more potent in vitro than the native GLP sequence, and its analogs are appreciably soluble in PBS. Throughout this book, GLP(7-36)-CEX amide has been denoted simply as GLP(7-36)-CEX with changes added to this nomenclature to signify related peptides.

4.3 Adding Dipeptides to the N-Terminus of F^7, GLP(8-36)-CEX

$G^5G^6F^7$, GLP(8-36)-CEX (Class 1) and OH-$G^5G^6F^7$, GLP(8-36)-CEX (Class 2) were synthesized to determine if either of the nucleophiles (amine or hydroxyl) could cleave the amide bond by 2,5-diketopiperazine or 2,5-diketomorpholine formation respectively, and thus regenerate the F^7, GLP(8-36)-CEX. The two compounds are shown below (Fig. 4.4a). The schematic synthesis of the Class 2 prodrug is represented in Fig. 4.4b. All subsequent Class II prodrugs were synthesized by analogous procedures.

The analyses of the two peptides after incubation in PBS (Fig. 4.4a; Table 4.3) showed that neither the amine, nor the hydroxyl nucleophile could cleave the amide bond. The structures of the peptides in Table 4.3 are also illustrated in Appendix IV (along with the relevant stereochemistry).

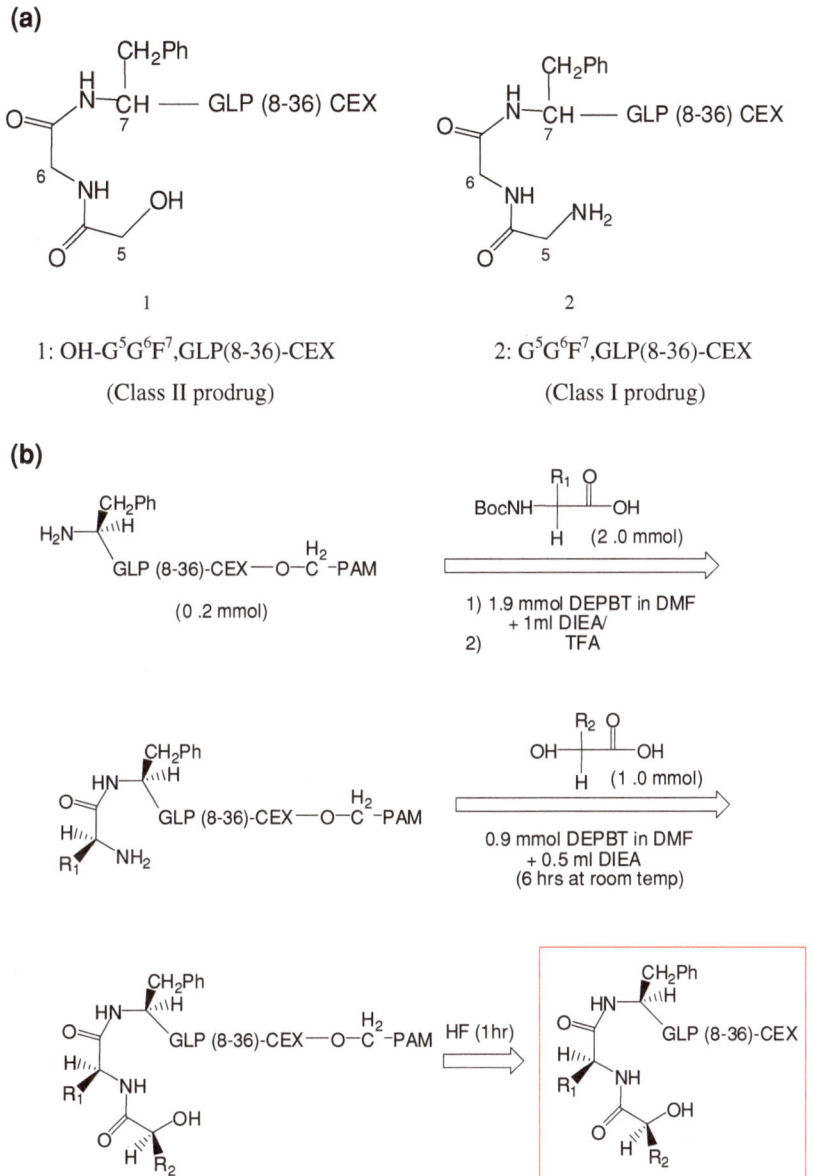

Fig. 4.4 a Dipeptide extended F⁷, GLP(8-36)-CEX. **b** Schematic synthesis of Class II prodrugs

Representative HPLC analyses of $G^5G^6F^7$, GLP(8-36)-CEX dissolved in PBS before and after treatment at 100 °C for 2 h is shown in Fig. 4.5. This analysis was done with the crude peptide. The amide bond did not cleave and the DKP was not generated. This was reflected by the absence of a shift of the HPLC peak in the

Table 4.3 Attempted cleavage of dipeptides attached to the N-terminus of F^7, GLP(8-36)-CEX

Serial no	Peptide (Xaa^{1}Yaa^2-F^{7a}-GLP-CEX)	Rate of cleavage at 100 °C	Rate of cleavage at 37 °C
1	GG (Class 1)	No cleavage	No cleavage
2	HO-GG (Class 2)	No cleavage	No cleavage

[a] In all these cases, there is phenylalanine in the first position (phenylalanyl leaving group) of GLP

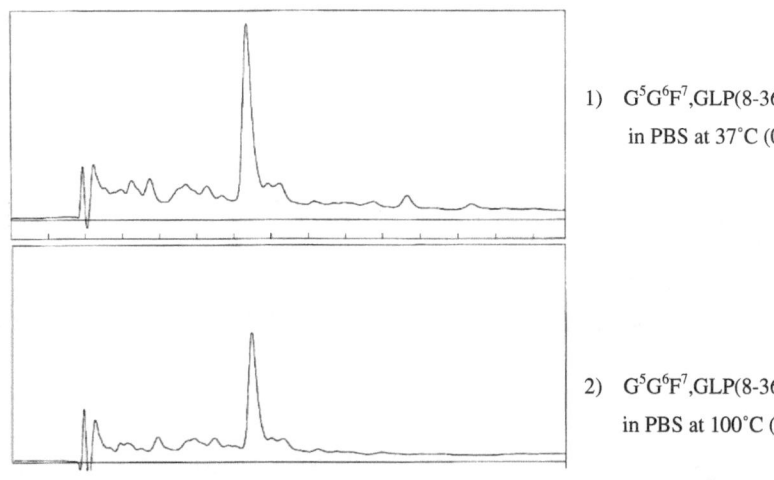

1) $G^5G^6F^7$,GLP(8-36)-CEX
   in PBS at 37°C (0 min)

2) $G^5G^6F^7$,GLP(8-36)-CEX
   in PBS at 100°C (120 min)

Fig. 4.5 HPLC analysis of $G^5G^6F^7$, GLP(8-36)-CEX

Table 4.4 Attempted cleavage of dipeptides added to the N-terminus of G^7, GLP(8-36)CEX

Serial no	Peptide (Xaa^{1}Yaa^2-G^{7a}-GLP-CEX)	Rate of cleavage at 100 °C	Rate of cleavage at 37 °C
3	GG (Class 1)	No cleavage	No cleavage
4	HO-GG (Class 2)	No cleavage	No cleavage

[a] In all these cases, there is glycine in the first position (glycyl leaving group) of GLP

elution profile of the prodrug to that of the dipeptide shortened peptide drug F^7, GLP(8-36)-CEX (not shown here) even when heated at 100 °C.

In Table 4.4, the analyses of two peptides with a glycine at the N-terminal position are shown. These peptides were tested to check the effect of the less bulky glycyl leaving group, as opposed to the previously studied phenylalanine in Table 4.3. The structures of the peptides in Table 4.3 are displayed in Appendix V.

Based on the results shown in Table 4.4, it was concluded that the amide bond is very difficult to cleave under physiological conditions either by 2,5-diketopiperazine (DKP) or 2,5-diketomorpholine (DMP) formation. This is independent of the nucleophiles and leaving groups tested, even under elevated temperature. The focus of this study then moved to esters which were anticipated to be easier to cleave as compared to the amides.

4.4 Depsipeptides and Esters

Depsipeptides were synthesized through addition of dipeptides to the hydroxyl group at the N-terminus of another peptide via an ester linkage (Fig. 3.2). The coupling procedures are described in the experimental section and they proved highly effective. As before, dipeptides of differential tendency for intramolecular cyclization (diketopiperazine formation) and release of the parent drug (N-terminal hydroxyl peptide) were studied. While adding the dipeptide, both Class 3 (an amine nucleophile cleaving an ester bond) and Class 4 (a hydroxyl nucleophile cleaving an ester bond—Fig. 3.3) compounds were prepared and studied.

The initial work began with imidazole-lactic acid (OH-His) as the terminal amino acid and subsequent leaving group. The HO-His[7], GLP(8-37) was synthesized. It was a full agonist with 25 % the potency of native GLP-1. There were several attempts to add a dipeptide to the α hydroxyl group of HO-His[7], GLP(8-37) and test for the cleavage of the new ester bond. However, there was a problem in selectively acylating the α-hydroxyl group of this peptide. As "protected" imidazole-lactic acid was not commercially available, we decided initially to work with the compound where the imidazole group was unprotected. As a result, the "unprotected imidazole" was inadvertently acylated along with the hydroxyl group. A synthetic scheme for the preparation of HO-His[7], GLP(8-37), its subsequent acylation and the formation of multiple side-products is depicted in Appendix VIII.

Consequently, Phenyllactic acid (OH-phenylalanine) was chosen in this initial test as it was readily available, and the return to a protected HO-His would be conducted if the work with HO-Phe proved successful.

4.5 Adding Dipeptides to the OH Terminus
of HO-F[7], GLP(8-36)-CEX

The HO-F[7], GLP(8-36)-CEX peptide was synthesized from the GLP(8-36)-CEX PAM resin (the synthetic scheme is represented in Fig. 4.6a) and served as the parent drug for these experiments. This peptide has an EC_{50} value of 0.008 nM (gold color in Fig. 4.6b), while native GLP has an EC_{50} value of 0.011 nM. Therefore, the HO-F[7], GLP(8-36)-CEX was found to be at least as potent as the native GLP-1. This is a most important and a somewhat unexpected observation that the histidine can be substituted by phenyllactic acid. This warranted additional

(a)

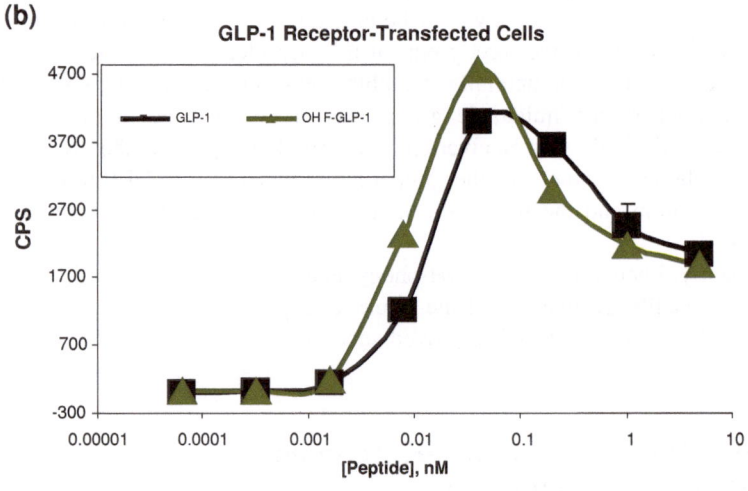

Fig. 4.6 **a** Schematic synthesis of HO-F^7, GLP(8-36)-CEX. **b** Bioassay results of HO-F^7, GLP(8-36)-CEX

validation. Subsequent experiments yielded identical results. The possibility exists that this result is a function of the CEX extension in the HO-F^7, GLP(8-36)-CEX. In this regard, it has been reported that the C-terminal region of exendin-4 increases its affinity to the GLP receptor [2].

The α hydroxyl group of HO-F^7, GLP(8-36)-CEX was acylated and both Class 3 (an amine nucleophile cleaving an ester bond; Fig. 4.7a) and Class 4 (a hydroxyl nucleophile cleaving an ester bond; Fig. 4.7(b) compounds were prepared and studied. Along with the 2,5-diketopiperazine (amine nucleophile) or 2,5-diketomorpholine (hydroxyl nucleophile), the biologically active HO-F^7, GLP(8-36)-CEX is released as shown in Fig. 4.7c.

To explore the possible formation of 2,5-diketopiperazine (DKP), or 2,5-diketomorpholine (DMP), and the simultaneous generation of HO-F^7, GLP(8-36)-CEX, the prodrug was incubated in PBS buffer at 37 °C for approximately a week. The peptide was also heated at 100 °C to determine the susceptibility to cleavage (Table 4.5). The structures of the peptides in Table 4.5 are illustrated in Appendix VI.

The first two peptides shown in Table 4.5 (Peptides 1 and 2) were analyzed to determine if either of the nucleophiles (amine or hydroxyl) could cleave the ester bond by DKP or DMP formation respectively. The results validated our expectation that the esters are more susceptible to cleavage than the amides. The relatively slower rate of cleavage of peptides 3, 4, 8, 9, and 10 seem to suggest that the bulk of the dipeptide extensions (Phe or Val) greatly affect the rate of dissociation of the prodrug. While there is virtually no difference between a methyl and a hydrogen side chain (5–7), the presence of an isopropyl group (β branching) greatly attenuates the rate of ester cleavage as evidenced in peptides 3, 4, and 10. Peptide 10 dissociates faster than peptide 3 as the amine is a stronger nucleophile as compared to the hydroxyl, a point corroborated by peptides 1 and 2 as well. Finally, the large difference in the $t_{1/2}$ between peptides 8 and 9 indicates that the side chain interactions in the dipeptide (R_1 and R_2 in Fig. 4.7c) play an important role as well. This same effect is shown by peptides 3 and 4. In both instances, the L,D-dipeptide diastereoisomer cleaves faster than the corresponding L,L-diastereoisomer.

Peptides 3–9 are represented as shown in Fig. 4.8a. The observed half-life of the prodrug increases as the substituent X gets bulkier (3, 4, 8, 9, and 10 in Table 4.5). This is probably because the transition state (TS) is more sterically hindered as the size of X increases. Consequently, the energy of activation increases thereby increasing the half-life of the prodrug (refer to Fig. 4.12).

This is especially exemplified for the two pairs of stereoisomers (3 and 4; 8 and 9). Here, by changing the stereochemistry of a single amino acid in the C-terminus of the dipeptide extension, a huge difference in rate was observed. The steric hindrance of the corresponding TSs is far greater in a L,L-dipeptide diastereoisomer (Analog 1 in Fig. 4.8a) as compared to a L,D-diastereoisomer (Analog 2 in Fig. 4.8a). Hence, 3 has a greater $t_{1/2}$ as compared to 4, and 9 has a $t_{1/2}$ greater than 10.

Compound **10** is F^{5}V^6-O-F^7, GLP(8-36)-CEX (Fig. 4.8b). This has a $t_{1/2}$ of 64 h and represents the longest duration peptidic prodrug to cleavage under physiological conditions, reported to date. The TS in this case is sizably hindered because of the steric interaction between the phenylalanyl and the valyl side chain.

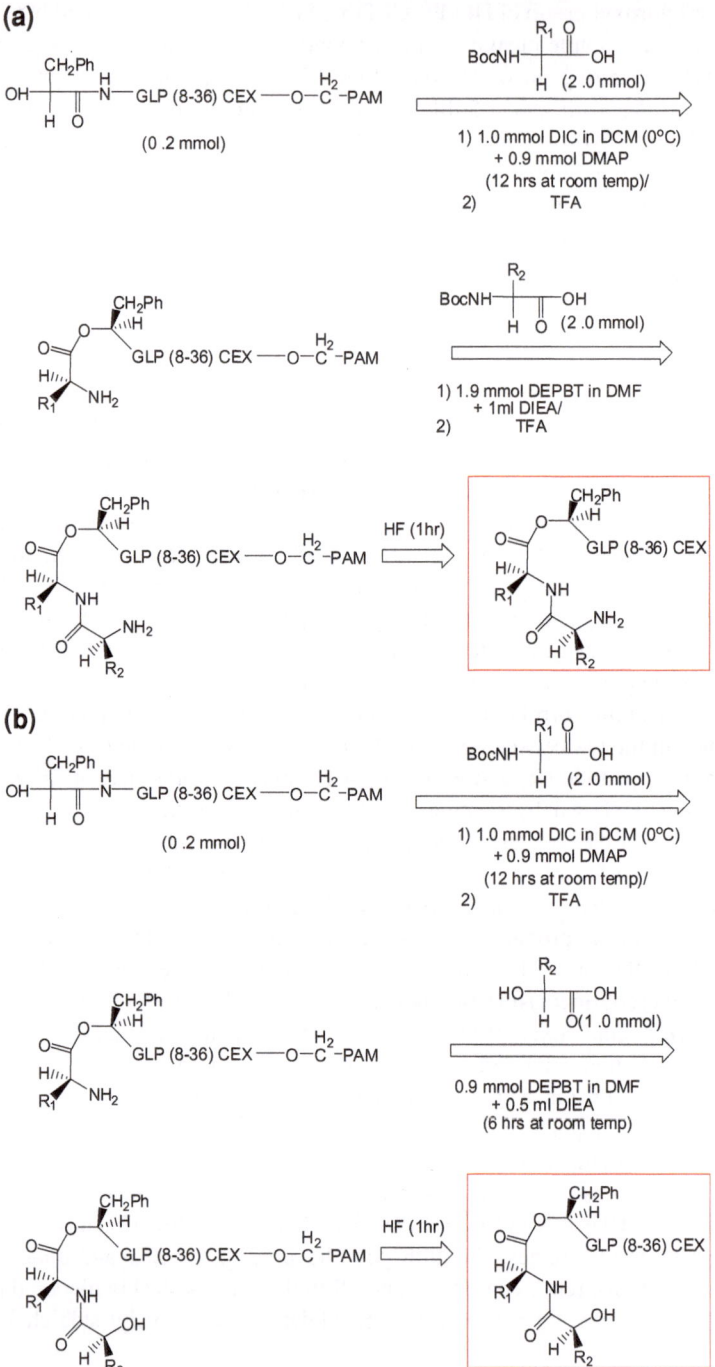

Fig. 4.7 a Schematic synthesis of Class III prodrugs. **b** Schematic synthesis of Class IV prodrugs. **c** Cleavage of **a** amine cleaving ester (Class III) and **b** hydroxyl cleaving ester (Class IV) prodrugs

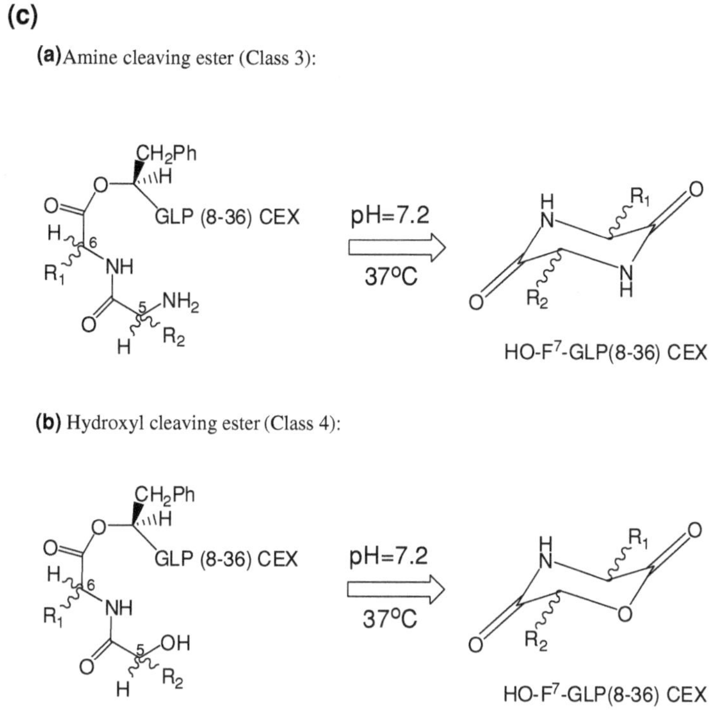

(c)

(a) Amine cleaving ester (Class 3):

pH=7.2
37°C

HO-F[7]-GLP(8-36) CEX

(b) Hydroxyl cleaving ester (Class 4):

pH=7.2
37°C

HO-F[7]-GLP(8-36) CEX

Fig. 4.7 continued

Table 4.5 Cleavage of dipeptide extended HO-F[7], GLP(8 36)-CEX

Serial no	Peptide (Yaa[1]Xaa[2]-O-Phe[7]-GLP-8-36-CEX)	$t_{1/2}^{a}$ of cleavage at 37 °C (h)
1	Gly-Gly (Class 3)	0.87
2	HO-Gly-Gly (Class 4)	1.13
3	HO-Phe-Val (Class 4)	No cleavage
4	HO-Phe-dVal (Class 4)	50.58
5	HO-Phe-Ala (Class 4)	2.83
6	HO-Phe-dAla (Class 4)	2.41
7	HO-Phe-Gly (Class 4)	2.46
8	HO-Phe-Phe (Class 4)	33.31
9	HO-Phe-dPhe (Class 4)	7.65
10	Phe-Val (Class 3)	64.0

[a] $t_{1/2}$ is the time required for 50 % release of the HO-Phe-GLP8-36-CEX at 37 °C in PBS (pH 7.2). It is calculated by a standard first order reaction plot. All these esters, including #3 cleaved at 100 °C. In all these cases, there is HO-Phe in the first position (hydroxy phenylalanyl leaving group) of GLP

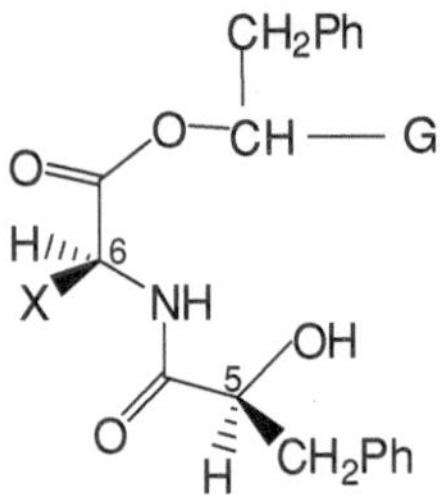

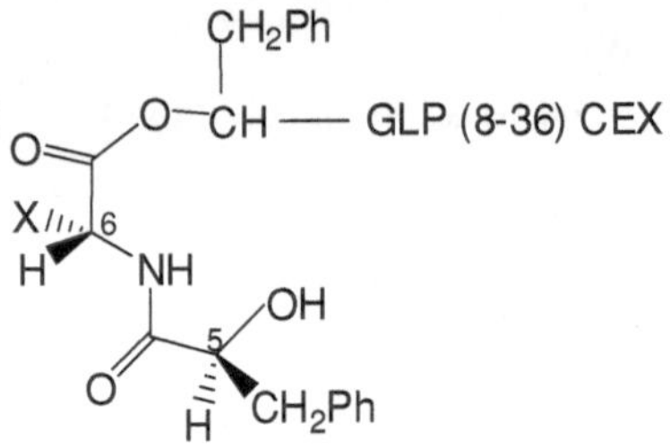

1

1: L,L-dipeptide extension

2

2: L,D-dipeptide extension

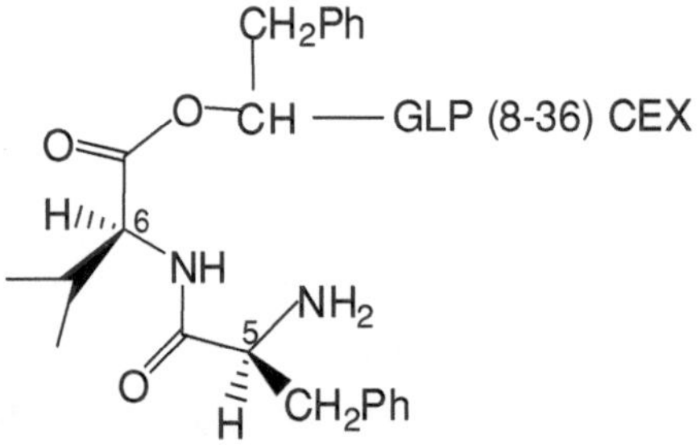

Fig. 4.8 **a** HO-F^{5}X^6-O-F^7, GLP(8-36)-CEX structures. 1: L,L-dipeptide extension; 2: L,D-dipeptide extension. **b** F^{5}V^6-O-F^7, GLP(8-36)-CEX ($t_{1/2}$ = 64 h)

The experimental data supporting the cleavage of compound **10** (Fig. 4.8b) (4264 Da) to form the parent drug HO-F^7, GLP(8-36)-CEX (4019 Da) is shown in Fig. 4.9. Since the reaction studied is an intramolecular cyclization, it is assumed that the presence of other impurities in the sample does not affect the rate of internal dissociation. In addition, both our molecules of interest (the prodrug and the drug) have been followed with a MALDI analysis even in the midst of contaminating material.

A second set of analogous depsipeptides were also studied to further examine the effects of the dipeptide extension structure upon the rate of cleavage (Table 4.6). Their synthesis follows the same pattern as outlined in Fig. 4.7a, b. The major difference is that in this case, the R_2 or Y site is a less hindered hydrogen (glycyl residue) instead of a bulky phenylalanyl as used previously (refer to Figs. 4.7c, 4.10). In this case, a glycolic acid (OH-glycine) or a glycine was used at the terminal end of the peptidic prodrug. The structures of the peptides in Table 4.6 are illustrated in Appendix VII.

The analysis was completed in the same manner as explained previously. The first two peptides shown in Table 4.5 (Peptides 1 and 2) were analyzed to determine if either of the nucleophiles (amine or hydroxyl) could cleave the ester bond

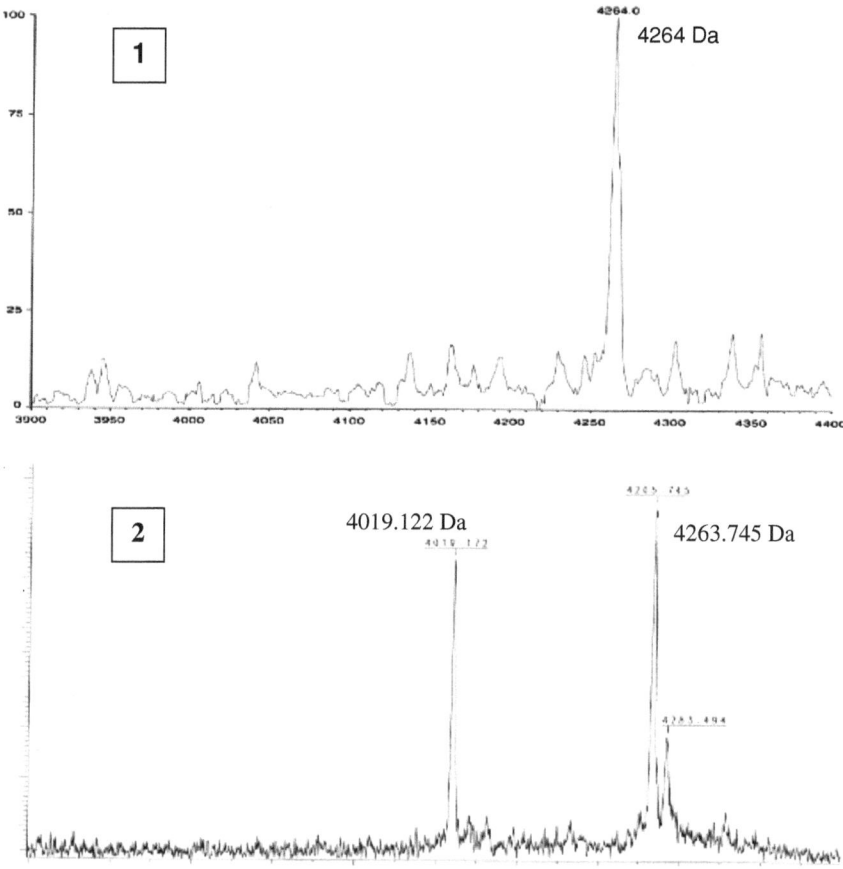

Fig. 4.9 Cleavage of F^{5}V^6-O-F^7, GLP(8-36)-CEX to form HO-F^7, GLP(8-36)-CEX. **a** Phe-Val-O-F^7-GLP8-36 CEX (4264 Da) in PBS at 37°C at 0 min. **b** Phe-Val-O-F^7-GLP8-36 CEX (4264 Da) cleaving to the parent HO-F^7-GLP 8-36 CEX (4019 Da) after incubation in PBS at 37°C for 64 h

by 2,5-diketopiperazine or 2,5-diketomorpholine formation, respectively. These two peptides had been analyzed in Table 4.5 and were tested again to serve as a control. As expected, both of the nucleophiles studied could cleave the ester bond. This set of compounds (Analogs 2–9 in Table 4.6) can be represented as shown in Fig. 4.10.

These peptides are expected to cleave faster than in the previous case, as the steric bulk is removed (glycyl group in Fig. 4.10 as compared to the phenylalanyl in Fig. 4.8a). As the size of X becomes structurally bulkier, the TS gets increasingly crowded. Consequently, the energy of activation increases thereby increasing the half-life of the prodrug (refer to Fig. 4.12). This is what is seen in peptide 10 where X is an isopropyl (valine) group. When X is a tertiary butyl group (peptide 7), the compound does not cleave at all in PBS buffer (pH 7.2) at 37 °C

Table 4.6 Cleavage of ester bond in HO-G^5X^6-O-F^7, GLP(8-36)CEX

Serial no	Peptide (Yaa^{1}Xaa^2-O-Phe^7-GLP-8-36-CEX)	$t_{1/2}$ of cleavage at 37 °C (h)
1	Gly-Gly (Class 3)	0.87
2	HO-Gly-Gly (Class 4)	1.13
3	HO-Gly-Val (Class 4)	4.70
4	HO-Gly-dVal (Class 4)	5.13
5	HO-Gly-^{a}AIB (Class 4)	0.75
6	HO- Gly-^{b}PhG (Class 4)	0.49
7	HO-Gly-^{c}tBut (Class 4)	Did not cleave
8	HO-Gly-Phe (Class 4)	0.70
9	HO-Gly-dPhe (Class 4)	0.93
10	Gly-Val (Class 3)	20.38

$t_{1/2}$ is the time required for 50 % release of the HO-Phe-GLP8-36-CEX at 37 °C in PBS (pH 7.2). It is calculated by a standard first order reaction plot. All these esters cleaved at 100 °C. In all these cases, there is OH-Phe in the first position (hydroxy phenylalanyl leaving group) of GLP
^{a} *AIB* α-aminoisobutyric acid
^{b} *PhG* phenylglycine
^{c} *tBut* tertiary butyl

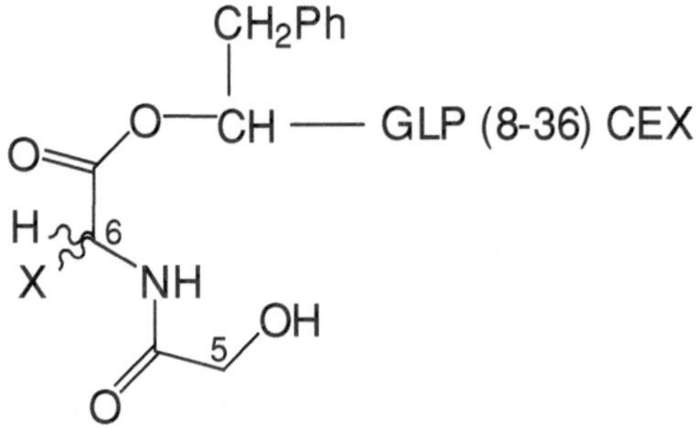

Fig. 4.10 HO-G^5X^6-O-F^7, GLP(8-36)CEX structures

(Fig. 4.11). In Table 4.6, the pairs of stereoisomers (Analogs 3 and 4; 8 and 9) dissociate at nearly the same rate. This is probably because the stereoisomers do not show a large difference in energy (purportedly because of the non-chiral nature of glycine). Hence, in this case, both the glycyl,L-dipeptide extension and the glycyl,D-dipeptide extension have a comparable energetic TS.

The observation that peptide 3 dissociates faster than peptide 10 seems to suggest that the hydroxyl is a stronger nucleophile as compared to the amine at a pH of 7.2 and 37 °C. One possible explanation for this anomaly might be that the pKa of the N-terminal amine (which is normally approximately 7) is slightly increased in peptide 10. As a result, at a pH of 7.2, this amine nucleophile would be disproportionately protonated and thus account for the slower rate of dissociation of the prodrug. This is just one possibility to explain the observed results, but additional study is needed to determine the exact basis of this observation.

1 | 2

1: G^{5}V^6-O-F^7,GLP(8-36)CEX ($t_{1/2}$=18 hrs)

2: HO-G^{5}tBuG^6-O-F^7,GLP(8-36)CEX (did not cleave at all)

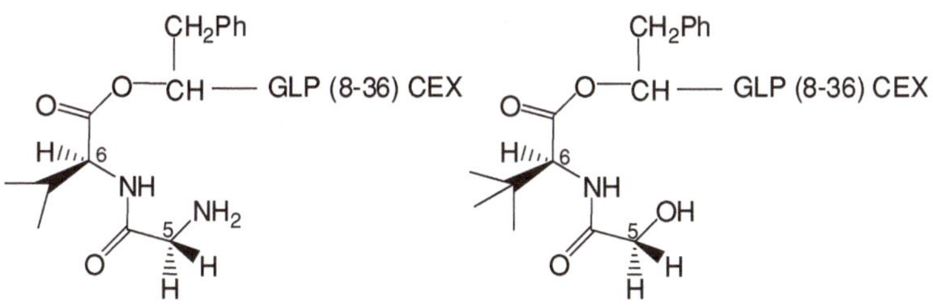

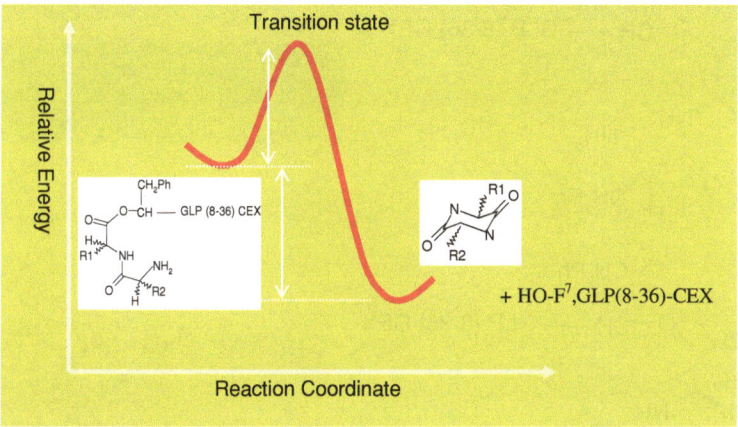

Fig. 4.11 Dipeptide extended HO-F^7, GLP(8-36)-CEX

Fig. 4.12 Transition state diagram

Table 4.7 Bioassay of select longer acting ester prodrugs

Serial no (color)	Peptide ($t_{1/2}$): (Yaa^{1}Xaa^2-O-Phe-GLP-8-36-CEX)	EC_{50} ± std. deviation (nM)	Percentage of potency with respect to parent
1 (red)	HO-F^7-GLP8-36 CEX (parent drug candidate)	0.028 ± 0.006	100 (parent drug)
2 (navy blue)	HO-Phe-dVal ($t_{1/2}$ = 50.5 h)	0.40 ± 0.06	7
3 (sky blue)	Phe-Val ($t_{1/2}$ = 64.0 h)	2.070 ± 0.41	1.3
4 (pink)	HO-Phe-Phe ($t_{1/2}$ = 33.3 h)	0.771 ± 0.249	3.6
5 (yellow)	Gly-Val ($t_{1/2}$ = 20.3 h)	0.230 ± 0.10	12

The color code refers to the respective peptides represented in Fig. 4.12. Percent potency calculations were done comparing the mean EC_{50} of the parent with that of the prodrug

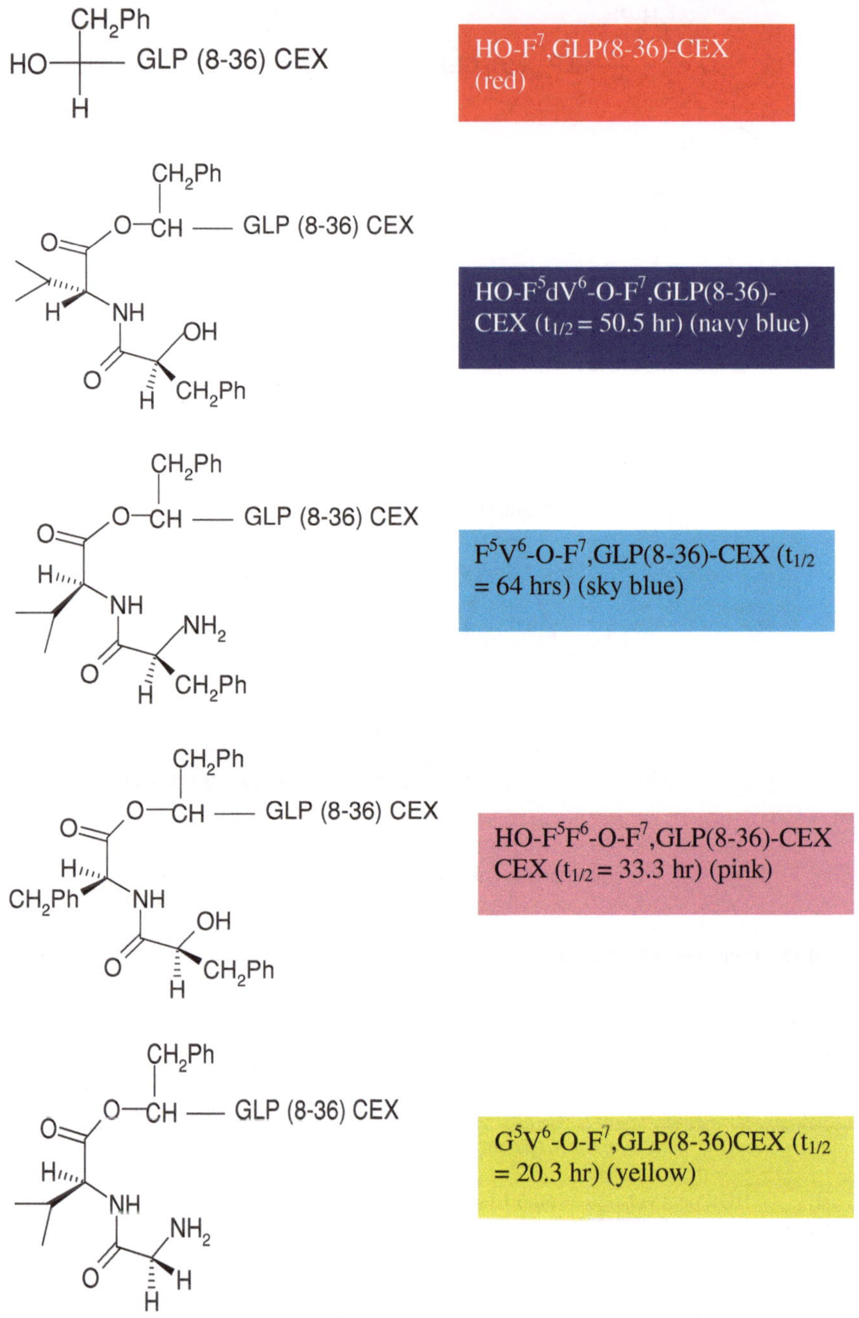

HO-F^7,GLP(8-36)-CEX
(red)

HO-F^{5}dV^6-O-F^7,GLP(8-36)-
CEX ($t_{1/2}$ = 50.5 hr) (navy blue)

F^{5}V^6-O-F^7,GLP(8-36)-CEX ($t_{1/2}$
= 64 hrs) (sky blue)

HO-F^{5}F^6-O-F^7,GLP(8-36)-CEX
CEX ($t_{1/2}$ = 33.3 hr) (pink)

G^{5}V^6-O-F^7,GLP(8-36)CEX ($t_{1/2}$
= 20.3 hr) (yellow)

Fig. 4.13 Structures of the peptides represented in Fig. 4.14 and Table 4.7

(a)

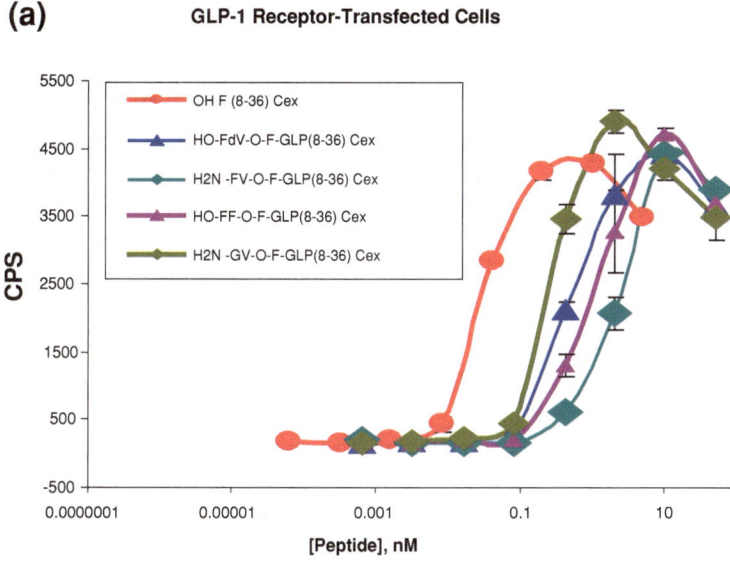

(b)

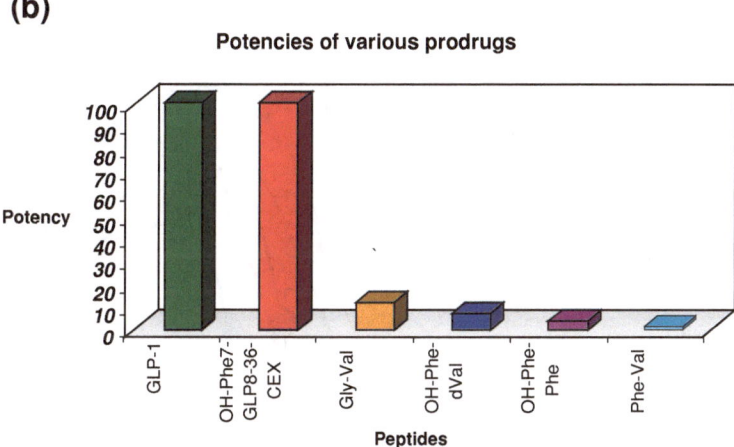

Fig. 4.14 a Bioassay results of select longer acting prodrugs. **b** Histogram showing the potency of various prodrugs

Thus, our results suggest that the structural nature of the side chain (especially β branching as evidenced in dipeptides containing valine and *tert*-butyl glycine); the stereochemistry and the pKa of the nucleophile serve an important role in determining the relative rate of cleavage. Several fast and slow acting ester prodrug candidates have been identified which were tested for their potency. As described earlier in our hypothesis, a prodrug should have minimal potency as compared to the drug, and its potency should increase consistent with the speed of cleavage of the dipeptide extension.

Table 4.8 Bioassay results that show the conversion of prodrugs to drugs

Serial no/ (color)	Peptide ($t_{1/2}$): (Yaa^{1}Xaa^2-O-Phe-GLP-8-36-CEX)	EC$_{50}$ ± std. deviation (nM)	Percentage of potency with respect to parent
Black	HO-F^7-GLP8-36-CEX (parent drug candidate)	0.023 ± 0.010	100 (parent drug)
5$^{(168\ h)}$ (red)^a	Gly-Val ($t_{1/2}$ = 20.3 h) incubated for 168 h	0.024 ± 0.012	95.8
4$^{(0\ h)}$ (navy blue)	HO-Phe-Phe ($t_{1/2}$ = 33.3 h) before incubation	0.612 ± 0.270	3.7
4$^{(24\ h)}$ (sky blue)	HO-Phe-Phe ($t_{1/2}$ = 33.3 h) incubated for 24 h	0.134 ± 0.050	17.2
5$^{(0\ h)}$ (pink)	Gly-Val ($t_{1/2}$ = 20.3 h) before incubation	0.211 ± 0.072	10.9
5$^{(24\ h)}$ (gold)	Gly-Val ($t_{1/2}$ = 20.3 h) incubated for 24 h	0.05 ± 0.020	46

$^{()}$ Time inside parenthesis refers to the time of incubation of the respective peptide. The color codes refer to the colors with which the respective peptides are represented in Fig. 4.13
^{a} Peptide 5 after incubation in PBS for a week (potency almost completely restored)

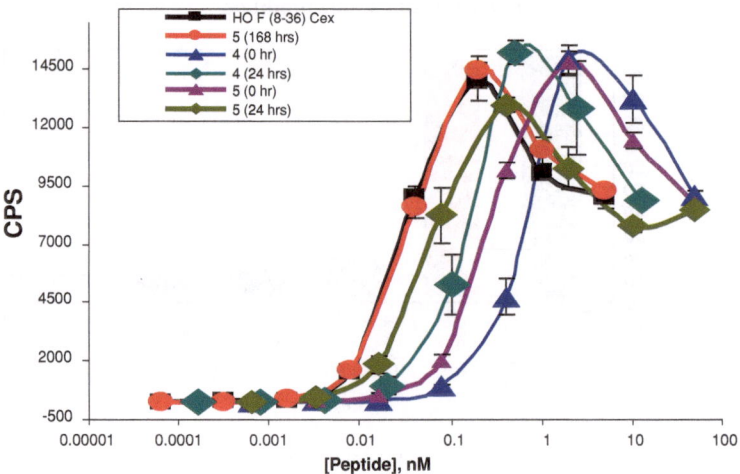

Fig. 4.15 Bioassay results that show the reversal of prodrugs to drugs

4.6 Bioassays of Selected Longer Acting Prodrug Candidates

Four of the longer duration prodrugs were chosen for further analysis in the biopotency tests. Luciferase-based bioassays were performed after purifying all these peptides by HPLC and confirming their masses by MALDI-MS analysis. Percent potency was calculated for purposes of comparing the mean EC$_{50}$ of the parent with that of the respective prodrug. The results are tabulated in Table 4.7 and then pictorially represented in Fig. 4.14a, b. The peptides tested in Table 4.7 are represented below in Fig. 4.13. The syntheses of these prodrugs are schematically shown in Appendix I.

These observations clearly show that addition of an ester linked dipeptide sequence to the terminal-hydroxyl group of the parent drug drastically reduces the potency of the drug. Two peptides were further analyzed to demonstrate that the prodrugs revert toward the potency of the parent drug following incubation for 24 h, in PBS (pH 7.2). The peptides chosen were analogs 4 and 5, based upon their physiologically relevant $t_{1/2}$ values being closest to 24 h (Table 4.8).

These experiments were performed using HPLC purified samples with the luciferase-based assay. Percent potency calculations were done comparing the mean EC_{50} of the parent with that of the prodrug. The results are tabulated in Table 4.8 and then pictorially represented in Fig. 4.15.

Prodrug 5 has its potency converted close to that of the parent drug after incubation for a week (shown in red). Both prodrugs display a higher potency after being incubated in PBS for 24 h due to gradual conversion to the parent drug. Thus, the definition of a prodrug has been substantiated by our data. Prodrug 4 exhibited a 17.2 % conversion (sky blue) after being incubated in PBS for 24 h. This is less than predicted by the calculated half-life of this prodrug (Table 4.7). The reason for this apparent disparity in the quantitative conversion rate of prodrug 4 has not yet been determined, though expected statistical variation in our bioassay results and the instability of the parent drug in the bioassay might account for this discrepancy. Future study of prodrug 4 will have multiple time points to help us understand this disparity.

References

1. Hamel AR et al (2004) J Peptide Res 63:147–154
2. Donnelly D (2003) J Biol Chem 278:10195–10200

Chapter 5
Conclusion

Abstract Clinical studies have revealed that GLP-1 therapy might be intrinsically safer than insulin therapy because of its glucose dependent action, thus eliminating the chances of hypoglycaemia. This book discusses the prodrug strategy as a means to extend the half life of GLP-1 which otherwise has a very short duration of action. An ideal prodrug should be stable to storage but must convert to the active drug under a specific set of conditions. These conditions for the activation of the prodrug will depend on its purpose and site of action. Detailed stability studies are important for the rigorous characterisation of promising prodrugs. However, a rapid screening helps in the identification of such prodrug candidates. The conditions described for the conversion here are a pH of 7.2 and temperature of 37 °C. These conditions were used for the cleavage of the prodrugs as they are physiologically invariant and can thus be translated into other peptidic drugs as well. The advantages of this strategy are discussed in this chapter. The biggest advantage is that the rate of conversion can be fine-tuned by selecting the structure of the pro-moieties. It is possible that the half lives of these prodrugs might be different when studied in vivo as compared to the in vitro experiments. This will probably be as a result of the action of the esterases and other enzymes, but unlikely a change in the chemical rate of conversion. The studies of the structure of additional pro-moieties and how they affect the rate of conversion will provide additional diversity in prodrug chemistry for future use.

Keywords Prodrug candidates · Physiologically invariant · Esterases · Dynamic range · Kinetics · Chemical conversion · Biological activity · Biological conversion

Clinical studies have revealed that GLP-1 therapy is an effective treatment for Type II diabetes. In addition, it might be intrinsically safer than insulin therapy because of its glucose dependent action, thus eliminating the chances of hypoglycaemia (see Chap. 1).

A. De, *Application of Peptide-Based Prodrug Chemistry in Drug Development*,   53
SpringerBriefs in Pharmaceutical Science & Drug Development,
DOI: 10.1007/978-1-4614-4875-4_5, © The Author(s) 2013

So far, no cases of hypoglycaemia have been reported using treatment with a GLP-related drug.

An ideal prodrug should be stable to storage but must convert to the active drug under a specific set of conditions. The desired set of conditions for the activation of the prodrug will depend on its purpose and site of action. Detailed stability studies are important for the rigorous characterisation of promising prodrugs. However, a rapid screening helps in the identification of such prodrug candidates. Such screening provides insights to the structure, stereochemistry and the site of attachment of the pro-moiety to the drug and how these affect the conversion of the inactive prodrug to the active drug. In the work presented here, the conditions for the conversion are a pH of 7.2 and temperature of 37 °C. These conditions were used for the cleavage of the prodrugs as they are physiologically invariant and can thus be translated into other peptidic drugs as well. The relative potency of prodrug candidates to the parent drug was determined through a receptor-based cell culture assay.

Prodrug strategies have often been implemented to improve the efficacy and safety of important drugs in different diseases [1]. In this report, the prodrug strategy has been used to increase the time action of the GLP analogue by extending the half life of the prodrug. Although a number of different pro-moieties could be used, the 'dipeptide pro-moiety or the α hydroxyl dipeptide analogue which dissociated by the formation of diketopiperazine (DKP) or diketomorpholine (DMP) along with the release of the active GLP analogue over a wide time range was chosen. Our approach has been subsequently published and acknowledged in the field [2–5].

Our strategy has a number of advantages. First, a number of structurally diverse amino acids (aliphatic, aromatic, acidic, basic, neutral) and α hydroxyl amino acids are commercially available. Second, given the natural nature of amino acids there is likely to be fewer safety and immunogenic concerns regarding their use as pro-moieties. In this context, it should also be noted that the native amino acid sequence was maintained as far as possible, and this should minimise potential adverse immunogenic responses. However, the magnitude and nature of such adverse reactions will need to be explored in vivo. Third, the amino acid properties and peptide synthetic chemistries are well established. Finally, and most importantly the chemistry of DKP formation allowed at least four points where the structure could be stereochemically controlled to refine the formation rate of the active GLP analogue. Prodrugs of varying half lives were synthesised by chemically modifying R1 and R2 in the context of different nucleophiles (Fig. 2.6). This generated the novel chemoreversible prodrugs of GLP, which converted to the active entity via the intramolecular cyclisation and subsequent release of 2,5-diketopiperazine (DKP) or 2,5-diketomorpholine (DMP). The large variance of half lives amongst the various prodrugs ranging from a $t_{1/2} = 0.5–64.0$ h underscores the great potential of this chemoreversible prodrug system in accurately tailoring the release of the parent drug under physiological conditions. The prodrug F^5V^6-O-F^7,GLP(8-36)-CEX with a $t_{1/2} = 64$ h is the longest acting chemically defined peptidic prodrug reported to date.

Four different varieties of prodrugs (Class 1, 2, 3 and 4) of GLP or biologically active GLP analogues were synthesised. Both the amide and the ester prodrugs were synthesised in good yield. In the beginning, the amide prodrugs were synthesised by adding a dipeptide to the amine group at the N-terminus. The cleavage of the amide bond of interest was attempted under physiological conditions with both an amine and a hydroxyl nucleophile (Tables 4.4 and 4.5). Different leaving groups were employed, with a histidine (Table 4.2), phenylalanine (Tables 4.3 and 4.4) and a glycine (Table 4.5) at the N-terminus of GLP analogues. For solubility purposes, we decided to work with a C-terminal modified GLP analogue, GLP-CEX.

We observed that the amide bond could not be dissociated by 2,5-diketopiperazine (DKP) or 2,5-diketomorpholine (DMP) formation under physiological conditions. In fact, the amide bond could not be cleaved even when heated at 100 °C at a pH = 7.2. Consequently, an amide prodrug could not be synthesised using this approach. As a result, our attention focused on synthesising ester prodrugs.

Two distinct classes of dipeptide esters were tested to study the cleavage of the ester bond. Class 3 variants employed an N-terminal amine nucleophile while Class 4 variants contained a hydroxyl nucleophile to cleave the ester bond (see Fig. 3.10). Selectively acylating the α-hydroxyl group of HO-H^7,GLP(8-36)-CEX proved to be very difficult in the absence of a "protected" imidazole peptide. Hence, we decided to work with phenyllactic acid and subsequently synthesised the HO-F^7,GLP(8-36)-CEX which was found to be at least as potent as the native GLP in repetitive analysis. The ester prodrugs were synthesised using this parent drug as the scaffold.

The esters proved to be much more labile than the corresponding amides and most of the ester prodrugs dissociated to give the active drug under physiological conditions along with the assumed formation of DKP or DMP. The degradation half lives of the respective prodrugs were calculated based on the first-order kinetics ($t_{1/2} = 693/k$) where 'k' is the first-order rate constant for the degradation of the prodrug. The disappearance of the prodrug and the appearance of the drug were both simultaneously analysed and calculated using HPLC. Generally, there was an excellent mass balance between the loss of the prodrug and the appearance of the drug. However in our results, only the estimated half life of the disappearance of the prodrug has been reported.

A sample kinetic profile reflecting the disappearance of the prodrug OH-Phe-dPhe-OF7-GLP(8-36)-CEX and the appearance of the parent drug HOF7-GLP(8-36)-CEX under physiological conditions is shown in Fig. 5.1. The slope of the first-order plot (blue diamonds) represents the dissociation constant of the prodrug from which the half life was calculated to be 2.8 h. The experiment was performed over several half lives and the recovery of the parent compound was more than 90 %.

Of all the ester prodrugs that were studied, only two did not cleave and they are represented in Fig. 5.2.

The most likely reason for this is the increased size of the R1 and R2 in compounds 1 and 2 (Fig. 5.2). The steric bulk of the isopropyl and tertiary butyl group crowds the transition state increasing the activation barrier for the

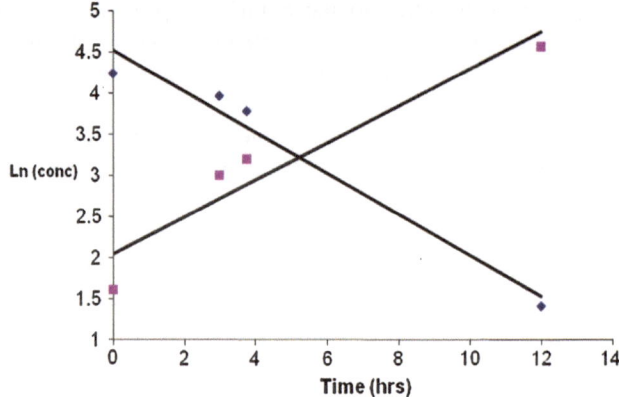

Blue diamonds represent the disappearance of the prodrug.
Pink squares represent the appearance of the drug.
The logarithm of the concentrations of the prodrug (blue) and the drug (pink) were plotted on the y
axis as a function of time on the x axis.

Fig. 5.1 Typical kinetic profile showing the disappearance of a prodrug and the appearance of a
drug

1: HO-F^{5}V^6-O-F^7,GLP(8-36)-CEX   2. HO-G^{5}tBut^6-O-F^7,GLP(8-36)CEX

Fig. 5.2 Ester prodrugs resistant to cleavage

intramolecular cyclisation. As a result, the cyclisation does not occur. When the
hydroxyl nucleophile in compound 1 was replaced with the amine, the corresponding F^{5}V^6-O-F^7,GLP(8-36)-CEX (Fig. 4.18b) cleaved with a long half life of
64.0 h. The presence of cleavage relative to the hydroxyl-peptide is because the
amine is a stronger nucleophile, as compared to the hydroxyl. Thus, it is seen that
the chemical stability of the ester bond is influenced by the structure of the amino
acid pro-moieties. The bulk of the side chain of the corresponding amino acids and
the strength of the nucleophile play a role in the cleavage of the ester prodrugs.
Additionally, stereochemistry of the two amino acids further influences the rate of
cleavage. These results are consistent with related observations in the field of
DKP-like prodrugs [6, 7].

Our primary goal was to synthesise longer acting prodrug candidates of GLP with a daily or even a weekly dosage. This is especially important as the longest acting DPP-IV resistant GLP analogue (Liraglutide) that is presently in Phase III trials has a biological half life of 10–14 h [8, 9]. Four prodrug candidates with varying half lives of 20.3, 33.3, 50.5 and 64.0 h were successfully synthesised (Fig. 4.23 and Appendix I). Each of these prodrug candidates should display extended duration of in vivo action and possibly constitute 'once a week' formulations, since this prolongation would be in addition to the natural time action of the parent drug.

The bio-assay we used is highly sensitive and reproducible. All the prodrugs have minimal inherent receptor potency relative to the parent drug (Table 4.8). The potency of G^5V^6-O-F^7,GLP(8-36)-CEX is nearly restored quantitatively (Fig. 5.3a), while that of HO-F^5F^6-O-F^7,GLP(8-36)-CEX is only partial after incubation in PBS at 37 °C for 24 h (Fig. 5.3b). The reason for latter observation is not immediately clear; and warrants further study.

In the overall context of things, we would have much preferred to have been able to cleave both the amide and ester bonds. This would provide the opportunity to utilise this prodrug chemistry at amine as well as hydroxyl containing drug candidates. One central disadvantage of ester prodrugs is the ubiquitous presence of esterases in vivo; and hence the effective half life of the ester prodrug might be much faster in blood then reported here.

In conclusion, our prodrugs were extremely stable when stored as a powder at 4 °C. They are soluble in physiological buffers, and exhibit differential conversion rates to the active drug under physiological conditions. This latter point is the central importance of my work. The results of this study clearly indicate that the cis-conformation of the dipeptide favours cleavage. Furthermore, that the overall structure (especially β branching as evidenced by the slow rates in pro-moieties containing valine), stereochemistry of the side chains, and the strength of the nucleophile play a profound role in affecting the rates of prodrug cleavage. It was also importantly shown that rate of conversion can be fine-tuned by selecting the structure of the pro-moieties. This is of utmost importance in the design of a properly functioning prodrug, as represented in Fig. 5.4. The dynamic range in rate of cleavage ranged from an hour to almost half a week. This provides appreciable opportunity for choosing from a wide spectrum of potential prodrugs, each varying in duration of action. These prodrugs with their tailored rate of conversions can provide a versatile lead for further optimisation of their pharmacological profiles.

It is envisioned that the half lives of these prodrugs might be different when studied in vivo than reported in our in vitro experiments. This will probably be as a result of the action of the esterases and other enzymes, but unlikely a change in the chemical rate of conversion. Hence, an in vivo characterisation of these prodrugs shall be necessary. Even if the prodrugs display the required half life in vivo, it might be required to slow their renal clearance (the renal threshold is 38,000 Da while the prodrugs have a molecular weight of $\sim 4{,}500$). This can be done by pegylating the prodrugs at appropriate residues.

(a)

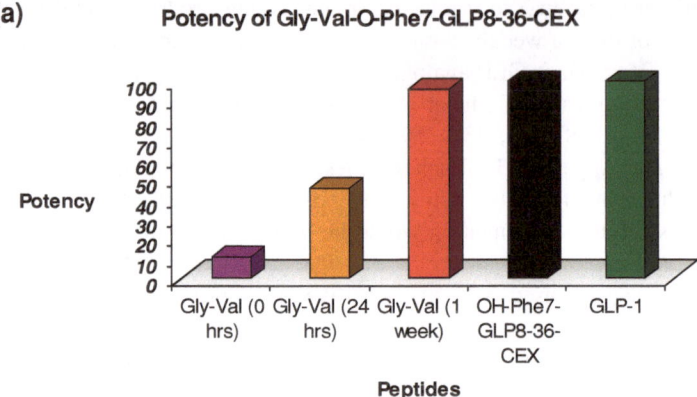

Green: GLP-1; Black: HO-^{7}GLP(8-36)-CEX
Red: G^{5}V^6-O-F^7,GLP(8-36)CEX after incubation in PBS for 1 week;
Yellow: G^{5}V^6-O-F^7,GLP(8-36)CEX after incubation in PBS for 24 hours;
Pink: G^{5}V^6-O-F^7,GLP(8-36)CEX in PBS at 0 hours;

(b)

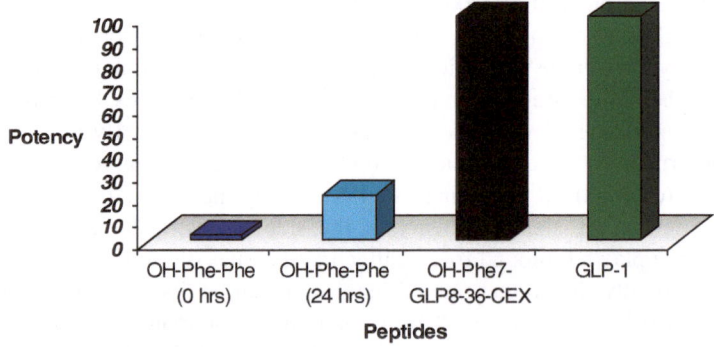

Green: GLP-1; Black: HO-^{7}GLP(8-36)-CEX
Light blue: HO-F^{5}F^6-O-F^7,GLP(8-36)-CEX after incubation in PBS for 24 hours;
Dark blue: HO-F^{5}F^6-O-F^7,GLP(8-36)-CEX in PBS at 0 hours;

Fig. 5.3 a Relative potency of G^{5}V^6-O-F^7,GLP(8-36)CEX, **b** Relative potency of HO-F^{5}F^6-O-F^7,GLP(8-36)-CEX

A place for additional investigation is the identification of methods to cleave amide prodrugs under physiological conditions. The amide prodrugs will presumably have an enhanced half life relative to that of the ester prodrugs. However, the increased flexibility of using amine-based drugs and their prodrugs to escape what might be selective degradation by esterases warrants further investigation.

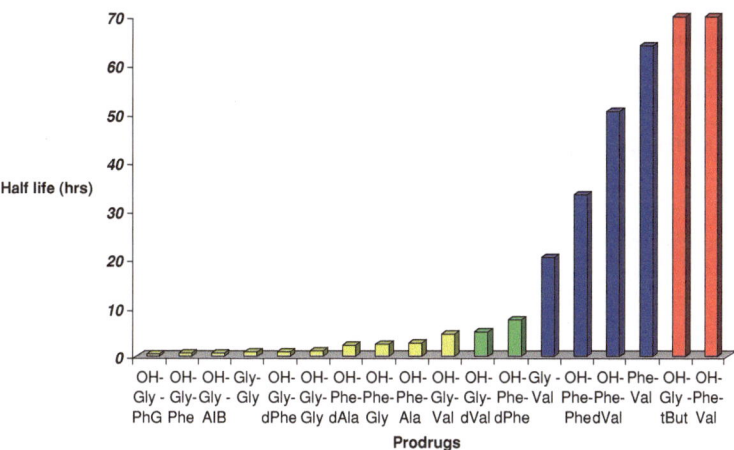

Fig. 5.4 Prodrugs with varying half lives

We have observed a number of prodrugs that could be potentially administered on a 'once a week' basis. Even longer acting prodrug candidates might be desired. Those will be prodrugs that can be administered on a 'once a month', or even in a 'once a year' formulation. However, it is worth noting that there are many other mechanisms beyond 'diketopiperazine formation' by which the prodrugs can be destroyed in vivo. Since peptides intrinsically decompose in vivo under physiological conditions, there is a limit set to the "time of action" of a specific peptide-based drug. However, our ability to use inverted stereochemistry of the dipeptide with no change in rate of chemical cleavage is a huge advantage in minimising enzymatic destruction of the prodrug element.

The studies of the structure of additional pro-moieties and how they affect the rate of conversion will provide additional diversity in prodrug chemistry for future use. This might then lead to a more effective structure-activity correlation study which can be subsequently used to synthesise prodrugs of optimal stability and pharmacokinetic profiles.

References

1. Denny WA (2001) Prodrug strategies in cancer therapy. Eur J Med Chem 36:577–595
2. De A, DiMarchi RD (2008) Investigation of the feasibility of an amide-based prodrug under physiological conditions. Int J Pept Res Ther 14:p255
3. De A, DiMarchi RD (2009) Synthesis and characterization of peptide hormone-based prodrugs. Pept Sci 92(4):310

4. De A, DiMarchi RD (2009) Synthesis & analysis of peptide hormone-based prodrugs. In: The proceedings of the 21st American peptide symposium, p 160
5. De A, DiMarchi RD (2010) Synthesis and characterization of ester-based prodrugs of glucagon-like peptide 1. Peptide Sci 94:448
6. Bundgard H, Moss J (1990) J Pharm Pharmacol 42:7–12
7. Borchardt RT, Cohen LA (1972) J Am Chem Soc 94:9166–9174
8. Knudsen LB et al (2000) J Med Chem 43:1664–1669
9. Bjerre K et al (2005) Diabetes 52(1):321–322

Appendix I
Schematic Synthesis of Longer Acting Prodrugs

(A) Synthesis of HO–F^{5}dV^6-O–F^7, GLP(8–36)-CEX

(B) Synthesis of F^5V^6–O–F^7, GLP(8–36)CEX

(C) Synthesis of HO–F^{5}F^6–O–F^7, GLP(8–36)-CEX

(D) Synthesis of G^5V^6–O–F^7, GLP(8–36)CEX

Appendix II
Structure of Peptides in Table A.1

Serial no	Structure
1.	
2.	

(continued)

A. De, *Application of Peptide-Based Prodrug Chemistry in Drug Development*,
SpringerBriefs in Pharmaceutical Science & Drug Development,
DOI: 10.1007/978-1-4614-4875-4, © The Author(s) 2013

(continued)

Serial no	Structure
3.	
4.	
5.	

(continued)

Serial no	Structure
6.	
7.	

Appendix III
Structure of Peptides in Table A.2

Serial no	Structure
1.	
2.	

A. De, *Application of Peptide-Based Prodrug Chemistry in Drug Development*,
SpringerBriefs in Pharmaceutical Science & Drug Development,
DOI: 10.1007/978-1-4614-4875-4, © The Author(s) 2013

Appendix IV
Structure of Peptides in Table A.3

Serial no	Structure
1.	
2.	

A. De, *Application of Peptide-Based Prodrug Chemistry in Drug Development*,   71
SpringerBriefs in Pharmaceutical Science & Drug Development,
DOI: 10.1007/978-1-4614-4875-4, © The Author(s) 2013

Appendix V
Structure of Peptides in Table A.4

Serial no	Structure
1.	
2.	

A. De, *Application of Peptide-Based Prodrug Chemistry in Drug Development*, 73
SpringerBriefs in Pharmaceutical Science & Drug Development,
DOI: 10.1007/978-1-4614-4875-4, © The Author(s) 2013

Appendix VI
Structure of Peptides in Table A.5

Serial no	Structure
1.	
2.	

(continued)

A. De, *Application of Peptide-Based Prodrug Chemistry in Drug Development*,
SpringerBriefs in Pharmaceutical Science & Drug Development,
DOI: 10.1007/978-1-4614-4875-4, © The Author(s) 2013

(continued)

Serial no	Structure
3.	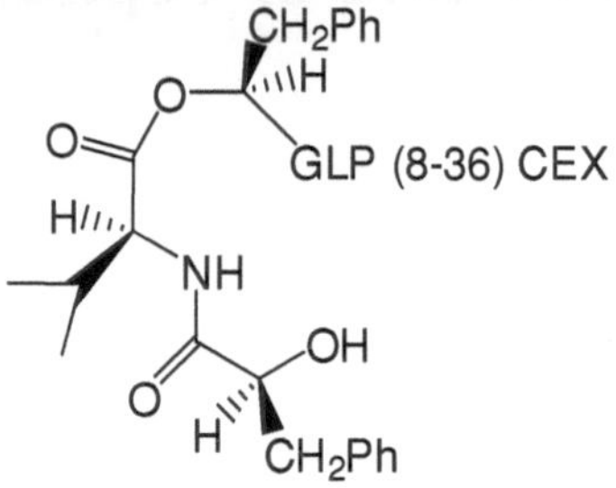
4.	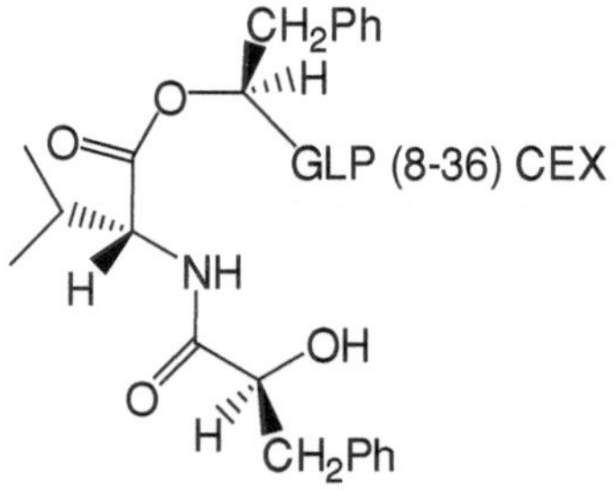
5.	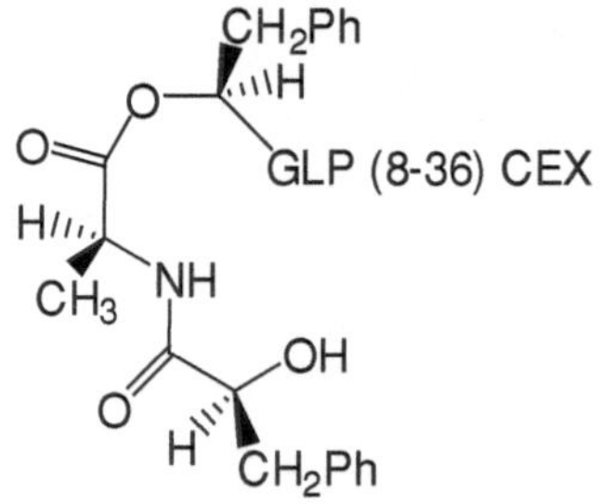
6.	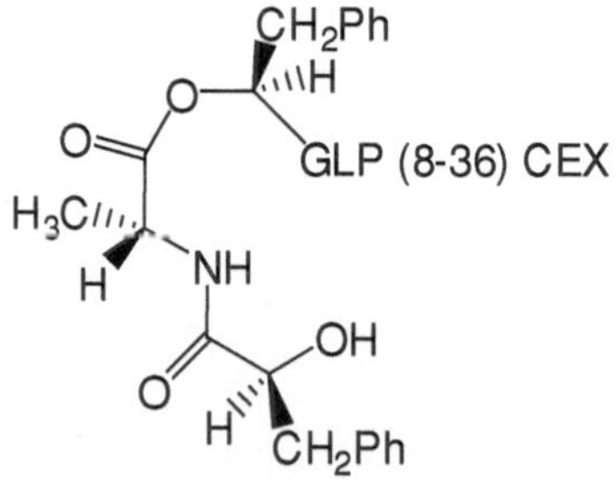

(continued)

(continued)

Serial no	Structure
7.	
8.	
9.	
10.	

Appendix VII
Structure of Peptides in Table A.6

Serial no	Structure
1.	
2.	

(continued)

A. De, *Application of Peptide-Based Prodrug Chemistry in Drug Development*,
SpringerBriefs in Pharmaceutical Science & Drug Development,
DOI: 10.1007/978-1-4614-4875-4, © The Author(s) 2013

(continued)

Serial no	Structure
3.	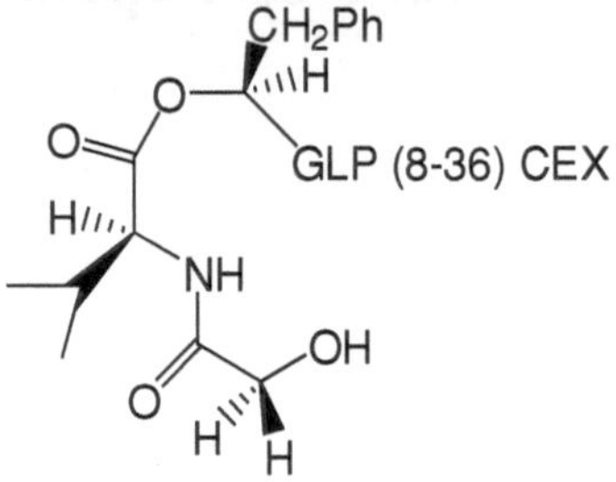
4.	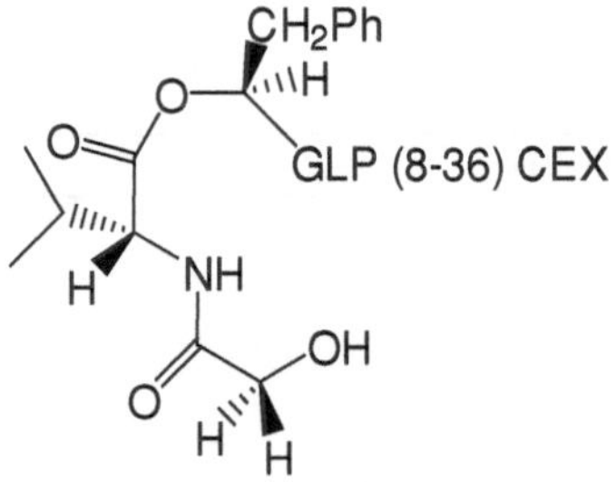
5.	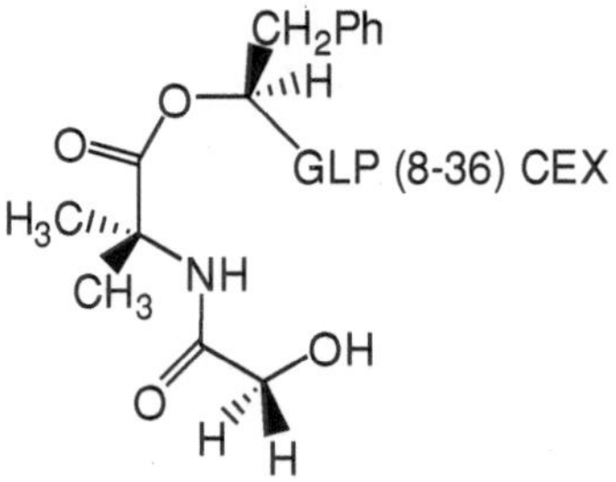
6.	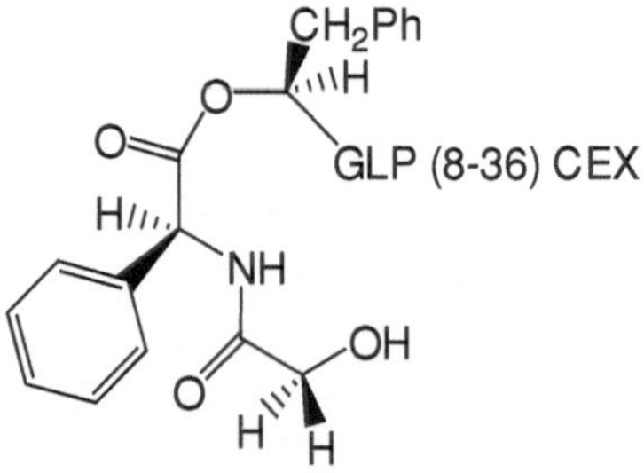

(continued)

(continued)

Serial no	Structure
7.	
8.	
9.	
10.	

Appendix VIII
Acylation of HO-His[7], GLP(8–37)

A. De, *Application of Peptide-Based Prodrug Chemistry in Drug Development*,
SpringerBriefs in Pharmaceutical Science & Drug Development,
DOI: 10.1007/978-1-4614-4875-4, © The Author(s) 2013

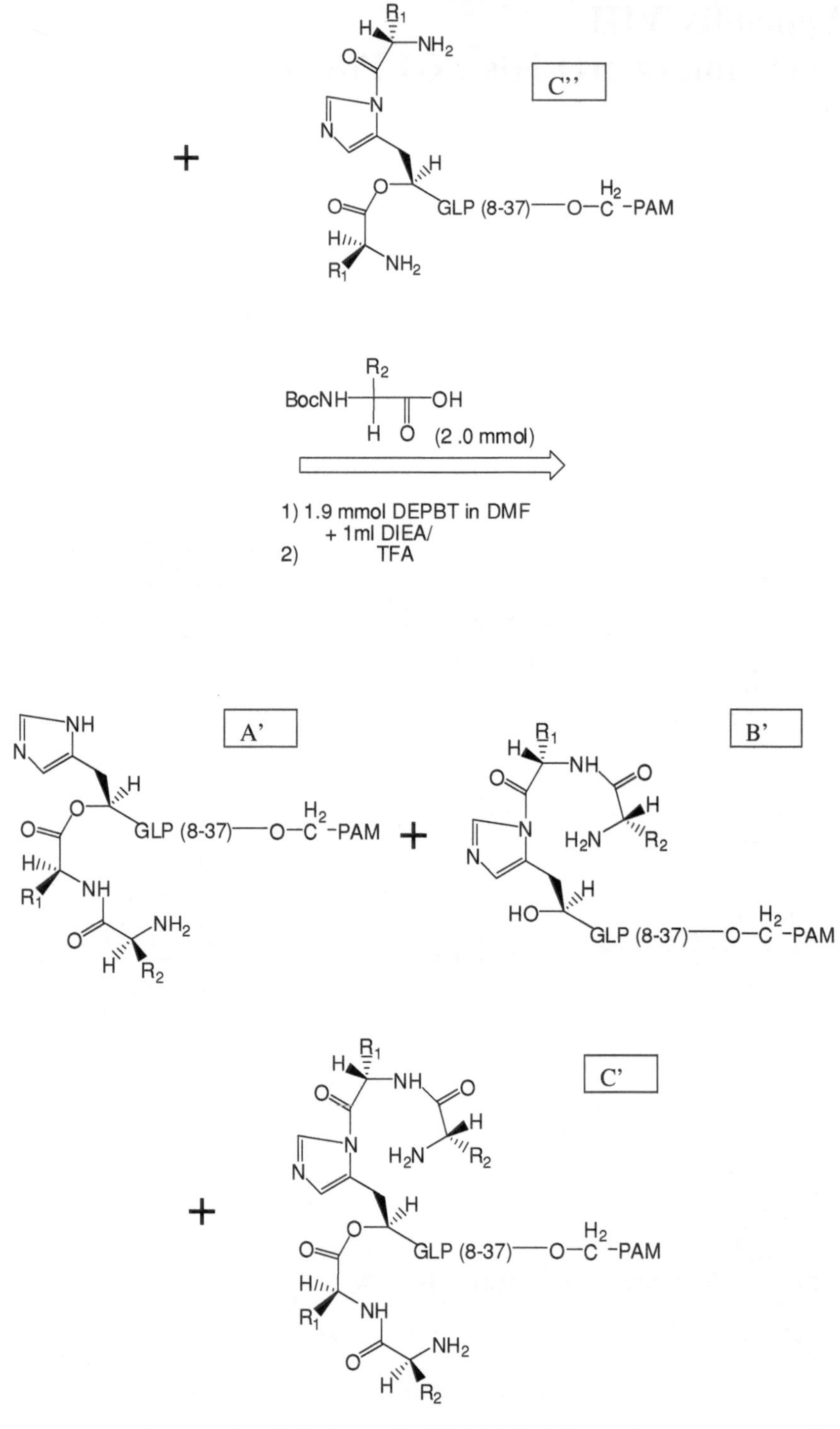
C''
BocNH–C(R2)(H)–C(=O)–OH (2 .0 mmol)
1) 1.9 mmol DEPBT in DMF
+ 1ml DIEA/
2) TFA
A'
B'
C'
GLP (8-37)—O–C(H2)-PAM

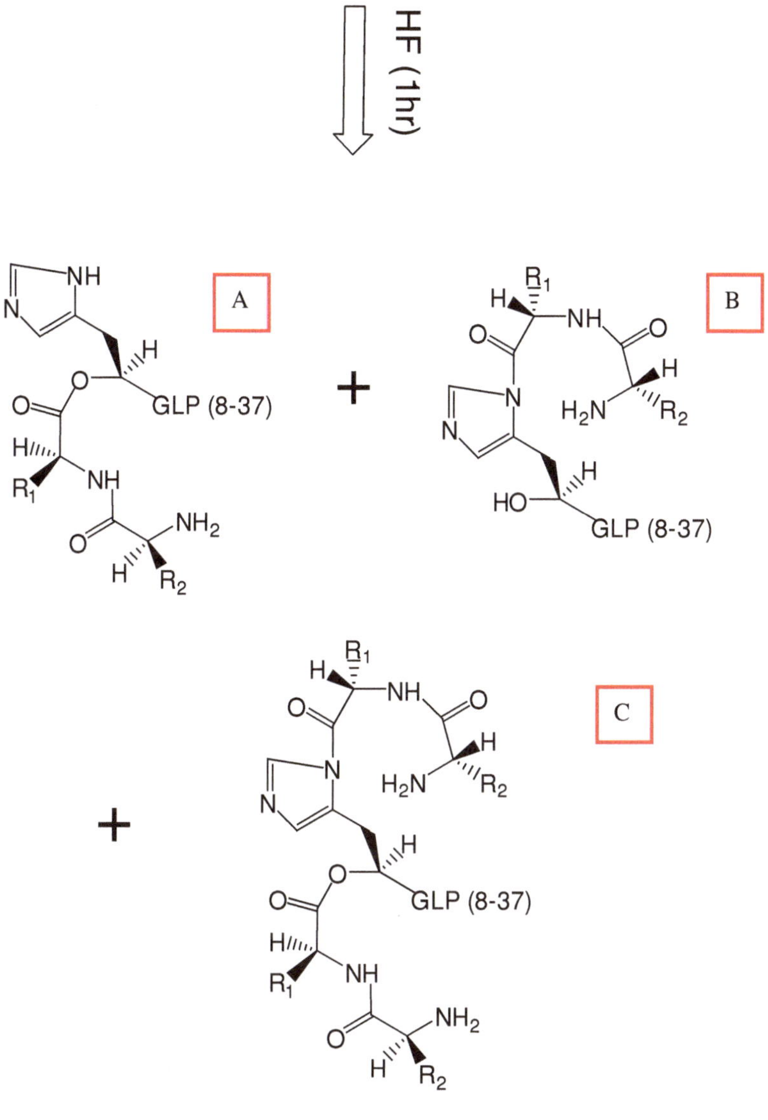
HF (1hr)
A
B
C
GLP (8-37)
GLP (8-37)
GLP (8-37)

Appendix IX
A Note on Nomenclature

A. Native GLP-1(7–37)-acid has been listed as GLP throughout this book. Its sequence is:

HAEGTFTSDVSSYLEGQAAKEFIAWLVKGRG-acid

B. F^7, GLP(8–37) refers to the GLP analog where the His at the 7th position has been replaced by Phe. The analog is:

FAEGTFTSDVSSYLEGQAAKEFIAWLVKGRG-acid

C. The GLP(7–36)-CEX amide has been designated as GLP(7–36)-CEX throughout this book. The CEX sequence (in red) is the C-terminal nine amino acids of exendin-4 where the C terminus is a serine amide. Its sequence is:

H^7AEGTFTSDVSSYLEGQAAKEFIAWLVKGRPSSGAPPPS-amide

Changes have been made to the nomenclature to signify analogous peptides. For example, the sequence of HO–F^7, GLP(8–36)-CEX is:

HO–F^7AEGTFTSDVSSYLEGQAAKEFIAWLVKGRPSSGAPPPS-amide

D. The amide prodrugs designated as $X^5Y^6H^7$, GLP(8–37) and $X^5Y^6F^7$, GLP(8–37) refers to the prodrugs synthesized on the native GLP and F^7, GLP(8–37) respectively. The sequence of $X^5Y^6H^7$, GLP(8–37) is shown below:

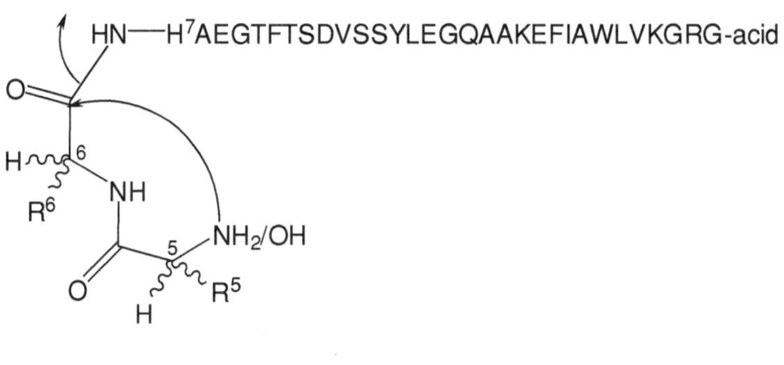

A. De, *Application of Peptide-Based Prodrug Chemistry in Drug Development*,
SpringerBriefs in Pharmaceutical Science & Drug Development,
DOI: 10.1007/978-1-4614-4875-4, © The Author(s) 2013

X⁵Y⁶H⁷AEGTFTSDVSSYLEGQAAKEFIAWLVKGRG-acid

The structure of such amide prodrugs (amine and hydroxyl nucleophile) is represented below:

E. The ester prodrugs are designated as X^5Y^6–O–F^7, GLP(8–36)-CEX with an amine nucleophile:

X⁵Y⁶F⁷AEGTFTSDVSSYLEGQAAKEFIAWLVKGRPSSGAPPPS-amide

and HO–X^5Y^6–O–F^7, GLP(8–36)-CEX with a hydroxyl nucleophile respectively.

HO–X⁵Y⁶-O-F⁷AEGTFTSDVSSYLEGQAAKEFIAWLVKGRPSSGAPP PS-amide

The structure of such ester prodrugs (amine and hydroxyl nucleophile) is represented below:

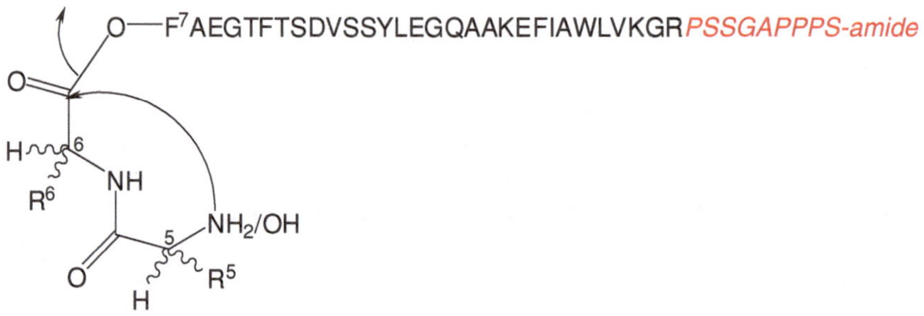

About the Author

Arnab De, M.A, M.Phil, is currently a PhD candidate at the Columbia University Medical Center. He completed his undergraduate education in Presidency College, Calcutta, India before coming to the Unites States for his higher education. He came to Indiana University, Bloomington where he worked with Prof. Richard DiMarchi (Standiford H. Cox Professor of Chemistry and the Linda & Jack Gill Chair in Biomolecular Sciences) to develop peptide-based prodrugs as therapeutics for diabetes. The work with Prof. DiMarchi resulted in two patents (licensed by Marcadia Biotech, recently acquired by Roche) and multiple publications in peer reviewed journals. He presented his findings in the American Peptide Symposium 2009 and received the Young Investigator's Award. He subsequently came to Columbia University where he is developing transgenic mice to serve as potential models for autoimmune diseases. He was invited by Carolyn J. Honour (Editorial Director, Biomedicine at Springer) to write this book in the Springer Brief series. The foreword has been written by Prof. Jean Martinez (Legion d'Honneur awarded by the French Republic and Chairman of European Peptide Society, 2002–2010).

A. De, *Application of Peptide-Based Prodrug Chemistry in Drug Development*,
SpringerBriefs in Pharmaceutical Science & Drug Development,
DOI: 10.1007/978-1-4614-4875-4, © The Author(s) 2013